Management Division

Core Programming Topic at the 2012 AIChE Spring Meeting and 8th Global Congress on Process Safety

Houston, Texas, USA
1-5 April 2012

ISBN: 978-1-62276-555-3

Printed from e-media with permission by:

Curran Associates, Inc.
57 Morehouse Lane
Red Hook, NY 12571

Printed by Curran Associates, Inc. (2012)

For permission requests, please contact AIChE
at the address below.

AIChE
3 Park Avenue
New York, NY 10016-5991

Phone: (203) 702-7660
Fax: (203) 775-5177

www.aiche.org

Additional copies of this publication are available from:

Curran Associates, Inc.
57 Morehouse Lane
Red Hook, NY 12571 USA
Phone: 845-758-0400
Fax: 845-758-2634
Email: curran@proceedings.com
Web: www.proceedings.com

TABLE OF CONTENTS

(4a) Equations of Change: How Patent Law Has Changed, and What Changes Are to Come

Jeffrey L. Wendt, The Wendt Firm PC, The Woodlands, TX

Patent law, particularly in the United States, has undergone many changes in the last 20 years, both in the courts and in the patent statute. My focus will be on what those changes have been, what changes are likely to come, and how these changes impact practicing chemical engineers. For example, 20 years ago the so-called "doctrine of equivalents" was widely used to ensnare infringers of patents. Now, the doctrine is almost an afterthought. My thoughts will touch on whether this is good or bad. Another change has been the term of patents. Before 1995, an issued patent had a term of 17 years from issue, even if the patent application had been filed many years before. Now, patents are granted for 20 years from the earliest filing date claimed as priority. Finally, I will touch on how a change to a "first to file" patent regime will make life different for engineer/inventors than the present "first to invent" system, particularly for small entities.

(4b) What Is Prior Art in the "First-to-File" Patent System Under the America Invents Act of 2011

Mark Rygiel, Sterne Kessler, Washington, DC and Anbar Khal, Sterne, Kessler, Goldstein & Fox PLLC, Washington, DC

On September 16, 2011, the Leahy-Smith America Invents Act was signed into law by President Obama. This law brings sweeping change to the current U.S. patent system, including transforming the current first-to-invent system into a first-inventor-to-file system. The first-to-file provisions of the Act take effect on March 16, 2013. These provisions substantially overhaul what acts are considered to be "prior art" and what acts are excepted. This presentation will provide an in-depth review of the new scope of prior art under the America Invents Act of 2011, focusing on what forms of disclosure prior to the effective filing date of a patent application will prevent the applicant from obtaining a patent in the United States. Case studies will be presented to illustrate the new prior art provisions, their exceptions, and how the new provisions compare to the existing prior art standards. The presentation will touch on how these new prior art provisions compare to the prior art rules of other developed countries having first-to-file systems.

Avoiding Common Mistakes in Licensing Agreements and Other Technology Agreements

By: Anthony P. Venturino, Esq. , Partner of Novak, Druce + Quigg, LLP Washington, DC

AIChE Spring Meeting - April 2, 2012 Houston, TX

The views presented here are those of the author and not attributable to his firm or its clients.

I. Technology agreements Can't live with them, can't live without them

Whether you consider technology agreements a help or a hindrance we need them to control the flow of technology. They set the ground rules so everyone involved knows how to handle technology being transferred and/or invented so there are no unwelcome surprises.

Thus, chemical engineers and managers need to know enough to know when they need an agreement or, how to perform if they are operating under an agreement.

In the course of performing their jobs chemical engineers and managers encounter many kinds of agreements having technology or other forms of intellectual property as a critical component. A few such agreements are as follows:

- Confidentiality Agreements
- Vendor Supplier Agreements
- Joint Development/Commercialization Agreements
- Exclusive Licenses / Non-exclusive Licenses
- Co-promotion Agreements
- Employee Agreements
- Settlement Agreements
- Joint Venture Agreements
- Contract R&D Agreements
- Consulting Agreements

II. Things can go wrong before, during and after the agreement

Here are some examples of where things can go wrong:

- Evaluation of the technology
- Selection of partners
- Conduct of the negotiations
- The agreement itself
- Implementing and servicing the agreement

Classic Licensee mistakes include:

- Royalties paid when licensed patent had expired
- Payments made after royalty terms had been satisfied
- Payments made under a trademark license when the trademark is not used

- Patents covered in a license no longer cover products that are currently made, yet royalty is paid against the patent coverage
- Royalties paid at the wrong rate, i.e. one higher or lower than provided Source:

Licensing – A Strategy for Profits, Edward P. White (1990)

Licensors also make mistakes. Here are some classics:

- Failure to take action if royalty payments cease when they should not, e.g., prior to expiration of the patents.
- Failure to monitor royalty payments, e.g. sudden reductions or failure to meet minimum royalties that could trigger a change from exclusive to nonexclusive or termination of the licenses.
- Failure to stay in technical contact with the licensee, with the result of licensee not interested in paying royalties, in maintaining quality, or keeping good records.

Licensing – A Strategy for Profits, Edward P. White (1990)

Before doing the deal.

- Know your business goals, strengths and weaknesses
- Know the available protection for the technology
- Have a realistic idea of the value of the technology
- Be open to creative technology agreements
- Do not let paranoia get in the way of a good deal
- Watch out for foreseeable legal and technical problems
- Admit when you are in over your head and get help
- Admit when you are in over your head and get help from a lawyer or other licensing professional (did I just say that?)

III. <u>Avoiding Common Traps</u>

A. <u>Define ownership and control of joint inventions of improvements</u>

Jointly developed invention often occur when smart people from different companies work together, for example while performing a joint R & D agreement or a variety of other agreements between companies. Joint inventions are jointly owned by US law unless there is a contract stating otherwise. This can cause problems.

If possible, have one party own the invention and, if necessary, grant the other party an exclusive field of use license. The following clause does this:

> In the event a joint invention arises during performance of this Agreement Company A owns the joint invention and controls patent prosecution and Company B gets an exclusive field of use license to the joint invention in the field of ______________.

B. <u>Maintain good records of what the agreement covers or when it expires</u>

The company, its legal department or a law firm should docket the agreement in a database. The docketing record should include: the type of agreement, the parties, the technology covered by the agreement, the agreement deadlines, the expiration date, the contact person and backup contact person, and have a miscellaneous portion for anything unusual.

Preferably an institutional person, for example, a docket clerk or contracts administrator, should be responsible for tracking notices in the docketing system and then relay the information to the business people. The more automatic you can make the timely conveying of such information, the better.

C. <u>Do not accidentally create an agreement</u>

The first time you negotiate with a party, especially if the other party sees this agreement as a one shot deal, be careful not to make records during the negotiations which unexpectedly bind you. Do not let an agreement to agree inadvertently become an agreement.

Summaries of the negotiations signed by both parties can lead to binding agreements even when a clause that might be important to you is missing. Let's take a hypothetical scenario where two parties bring their business people and lawyers to a meeting to negotiate a potential IP license. Mid-afternoon the attorney for the IP owner (potential licensor) had to leave to catch a plane. The potential licensor's business person (let's call him "Bill") had a later plane. The potential licensee kindly offered to take Bill to the airport later so he would not have to waste time in a boring airport. While Bill was by himself with the potential licensee's attorney and business people one of the business people suggested they draft and sign a summary of the day's discussions. They did this and then Bill caught his plane. The next day Bill's company was asked to sign an Assignment document to reflect the assignment they agreed to yesterday. Bill's response was "What assignment?" After a lawsuit the parties settled. The moral of the story is train the business people not to sign anything.

A deal can also inadvertently arise if the parties act like there is an agreement can create an agreement. This sometimes arises where the parties are in a rush to do joint development. While the deal is only a draft they start working together. Sometimes the unsigned draft becomes the *de facto* agreement.

The opposite can also happen. Sometimes you think you have an agreement but in reality you do not, or at least not all the clauses apply.

Take the case of *DataTreasury v. Wells Fargo,* 522 F.3d 1368 (Fed. Cir. 2008). In 2003 WMR e-Pin licensed a patent to Wells Fargo Services Corp. a subsidiary of Wells Fargo & Co. The license included an arbitration clause. In 2006 WMR e-Pin assigned 4 patents to DataTreasury. At least one of the 4 patents was a patent of the license. Afterwards DataTreasury sued Wells Fargo & Co. for patent infringement in court. Wells Fargo & Co. moved to dismiss or stay pending arbitration. Under 5th Circuit law, as applied by the

Federal Circuit, "to be enforceable, an arbitration clause must be in writing and signed by the party invoking it. *DataTreasury* at 1373, quoting *Westmorland v. Sadoux*, 299 F.3d 462, 465 (5th Cir. 2002). Neither Wells Fargo & Co. nor DataTreasury signed the original patent license. Thus, the court held these parties were not bound by the arbitration clause.

D. <u>Remember important clauses</u>

Clause checklists are important tools when reviewing a license to see if all the necessary clauses are present. However, use the right checklist for the type of agreement and the type of IP. Do not assume the clauses in your standard patent license belong in your new trademark license and *vice versa*. Unlike patents which protect the patent holder, trademarks protect the public by assuring the public a good labeled with a particular trademark will have the characteristics they expect. For example, a PIRELLI brand tire has certain handling properties. Trademark licenses without quality safeguards are termed naked licenses and invalidate the trademark rights.

A short checklist is provided in TABLE 1:

TABLE 1
Introductory Clauses
Definitions
Grant of Rights
Consideration/Compensation
Diligence
Product Quality Control
Advertising
3rd Party Infringement Suits
3rd Party Product Liability Suit
Licensor/ Licensee Suits
Termination
Continuing Obligations

A detailed set of clause checklists for a licensing agreement is available upon request from the author.

E. <u>Maintain control of the technology after making the deal</u>

Proprietary technology is a competitive advantage. If the licensor loses control of the technology he loses control of the competitive advantage. Thus, the licensor should determine the level of control he is willing to lose. Maybe the licensor thinks he can permit the licensee to sublicense because he trusts the judgment of licensee's president. What if the president leaves? Thus, the agreement preferably precludes sublicensing or assigning to competitors.

Another common way to lose control is for the licensee to be bought by the licensor's competitor. The Agreement should specify it may be assigned by ONE of the following:

1. Anyone;
2. Goes with the business;
3. Goes with the company;
4. Licensor maintains right of first refusal;
5. Licensor maintains right of first refusal which will not be unreasonably withheld;
6. No one.

F. Promote good communication after making the deal

Problems arise from misunderstandings, lack of thoroughness, carelessness, or failure to administer the agreement at all, i.e., no one in authority is responsible for administration.

The lawyer can structure the agreement to include mechanisms for communication between the licensor and licensee so little problems and misunderstandings do not become big ones. You cannot force the business people to communicate but the mechanisms provide fertile soil for communication. For example, the agreement should include a contact person and back-up contact person. The agreement can also mandate times for periodic meetings.

G. You need the equivalent of a pre-nuptial agreement

Agreements, like romance, can end badly. As with a newlywed couple, the enthusiastic partners to a new business relationship think nothing will go wrong. They can and they do.

Include termination provisions in the agreement to give each party a reasonable way to terminate the agreement. This includes planning for unexpected events, for example, bankruptcy or purchase of the other party. This also includes providing ways to terminate or renegotiate if the economics of the deal no longer make the current agreement viable.

IV. Organize A Licensing Department – A final word

Some companies only occasionally do a license or may be at the beginning of setting up a license project group. In these cases, the company might ask his or her lawyer for advice on how to organize the licensing project group.

The license project group will need the support of a top executive. Without this the license project group will generally not get the resources it needs.

Recognize the different functions to be performed. Then, divide or share these functions depending on available resources, business people, consultants or lawyers:

- Profit Center Group - optimizes the licensing in and licensing out to make money for the company
- Support Group - coordinates technical, legal and accounting aspects of tech transfer
- Administrative Group -handles accounting, record keeping and day to day details of maintaining the license.

V. Conclusion

Technology agreements help people work together while managing the intellectual property of their efforts. These agreements let everyone involved know their rights and responsibilities. Knowing your goal and having the appropriate agreement to help get to the goal and understanding the agreement can avoid nasty surprises. If you know your goals, technology and agreements you will be in a better position to turn that technology into profit.

(26a) Simplified Cost Estimation and Its Application to Decision Making: What Do You Really Need to Know?

John Eric Anderson, Core R+D, Engineering and Process Sciences, Dow Chemical, Midland, MI

Simplified Cost Estimation and its Application to Decision Making: What Do You Really Need to Know?

In an environment where new product and process concepts are actively encouraged, competition for both money and people will usually result. While it is true that this is a 'nice problem to have', this forces organizations to make decisions about where to expend limited resources. Valid economic forecasts are required but can be difficult to make in the formative stages of an idea's development since much of the required data is not available. However, in many cases, the impact that a particular unknown has on the economic results can be computed using reasonable ranges of input values and to clarify the economic picture. Case studies will be shared where decision-making was enhanced significantly using economic calculations that were sufficiently simplified that only a few days of effort were required.

(26b) Lessons Learned From SABIC New Processes Commercilization

Fuad Mosa, Sabic Technology Center, SABIC, Riyadh 11551, Saudi Arabia

During the Last 20 years, Sabic has experience within deferent affiliates many new or scaled-up processes where tremendous amount of resources have been devoted to operate and stabilize them. Commercial plant start-up of new process is the real evaluation of how much successful the critical development milestones were accomplished. All lesson learns from the start-up can be linked to the failure of estimating the required activities for each development step. Moreover, As the plant mainly design for steady state with specified margin, initial plant start-up can be considered the most challenging step which most likely not reflected in the design due to the difficulties to set all possible unknown scenario's which can be faced.

In fact, Development of new technology is including many critical milestones which can strongly affect the final plant set-up. Starting from process kinetics and selecting of the important elements for scale-up model, Building the initial concept of the process, choosing the important plant sections which required to be piloting till plant start-up are involving some challenging areas. The final evaluation of the completion for each step can be considered as an important element which can highly affecting the final plant performance.

Typically any new technology runs through an initial learning phases that could last upto many years depending on the complexity of the process. This paper will attempt to share with all sabic and provide the challenges faced in the journey of Many Process commercialization at Sadaf , Ibn Rushed , PK , United, ArRazi and ways to handle these challenges. This will be covered with a clear link between the important process commercialization milestones and start-up challenges faced which are very important for any team working on process development.

Presentation ID # 245903

Technical Support Benchmarking: A 15 Year Historical Perspective

Christopher F. Morrison, Ph.D.

Senior Research and Development Manager

Nalco, an Ecolab Company

Since 1996, the Nalco Downstream Petrochemicals Research and Development Team has utilized our Technical Support Request (TSR) database to document and resolve more than 4,000 customer petrochemical process issues that could not be resolved at either the local Nalco Sales Engineer or Regional Support level. This TSR database has withstood the test of time, and remains the primary workflow process for the delivery of "high level" support to our customers. Furthermore, this database serves as a rich knowledge repository, as well as a reliable tool to determine market trends and appropriate R&D resource allocation. This paper and associated presentation will discuss how technical support requests have trended with general market conditions, the more recent shift to gas cracking from naphtha cracking, regional shifts in ethylene capacity, unit shifts (e.g. increasing need for support in Primary Fractionators), as well as a statistical analysis of the composition of "typical" ethylene plant deposits.

Technical Support Requests (TSR's) are the primary workflow process for the delivery of "high level" Research and Development (R&D) support to our customers. If a customer problem cannot be resolved at either the Sales Engineer peer level or the regional Industry Technical Consultant (ITC) level, the problem comes into R&D. TSR's are used to solve problems (untreated or competitively held) or to address issues with existing Nalco treated programs. Over the past 16 years, the Nalco Downstream Petrochemicals R&D Team has utilized our TSR database to resolve and document 4,067 such customer petrochemical issues. For 9 of the last 15 years, the annual percent change in TSR's has directionally tracked with the annual percent change in global GDP. For 10 of the last 15 years, the annual percent change in TSR's has directionally tracked with the annual percent change in the total value of chemicals output. By and large, TSR activity is related to general market activity.

The TSR database is crucial to our operations since it's the primary workflow process for Marketing and R&D to deliver sales support. Perhaps more importantly, it is a central knowledge repository and a means for Nalco to share historical site information with our customers that they sometimes lose due to inevitable personnel changes at their customer sites. Furthermore, Nalco has utilized the TSR database to protect our intellectual property, ascertain market trends, and ultimately allocate resources towards issues important to our customers.

While the TSR database started out in the mid-1990's as a North America-centric tool, it was quickly embraced by our global team. As such, it has tracked with global trends in petrochemicals. For example, in 2011 Asia Pacific overtook Europe as the second most demanding region in terms of support behind North America. Recognizing this trend, Nalco redistributed its global Marketing and Sales force to Asia Pacific. Nalco also recognized the need for regional R&D support and constructed state of the art R&D facilities in Pune, India (2009) and Shanghai, China (2008).

Over a 16 year period, there were 1,865 TSR's covering the Olefins industry (excluding Butadiene). During this time frame, the most demanding application was Light Ends (24% of requests), followed by Dilution Steam (19%), Acid Gas Removal (17%), Compressors (16%), and Primary Fractionators (11%). The remaining 13% of requests were mainly around Furnaces, Pygas, and Hydrotreaters. The vast majority (83%) of requests were related to fouling issues, with corrosion (7%), foaming (4%), emulsions/water clarification (3%) and other (3%) rounding out the requests. While Light Ends accounted for the most requests (24%) over the 16 year period, it is worth noting that requests in both Light Ends and Acid Gas Removal have been trending downward, whereas requests for Compressors and Primary Fractionators have been trending upward. Light Ends (Actrene®) and Acid Gas Removal (Arsenal™) are Nalco programs that are particularly well refined and extensively deployed, and thus may account for decreasing support needs.

Since 1996, there have been 203 TSR's for Primary Fractionators. There were "spikes" in Primary Fractionator support in 1999, 2002, 2006, and 2009. With the exception of 1999, each "spike" in Primary Fractionator support was approximately 1 year after a significant increase in crude oil prices: in 2000 crude oil prices spiked 65% and late 2001 saw the beginning of the Afghanistan War; in 2005 crude oil prices spiked 33% (in large part due to Hurricanes Katrina and Rita); and in 2008 crude oil prices spiked 43% - surging to $140/barrel at one point. It is logical to assume that as crude oil prices increase, so too do naphtha prices, thereby compelling mixed crackers to entertain lighter feedstocks. The data we've gathered seems to suggest that there is a "delay" after customers shift to lighter feedstocks before they encounter problems that require high level technical support.

For the 393 TSR's related to Dilution Steam Systems, 37% involved the Dilution Steam Generators, 21% for Quench Water Settler Drums, 12% for Process Water Strippers, 11% for Quench Water Towers, 3% for Feed Saturator issues, 1% for DOX issues and 1% for Blowdown issues. In addition to these, 14% of the TSR's had multiple issues across the Dilution Steam System. Fouling in the Dilution Steam System accounted for the majority of the requests (61%), though corrosion (22%) and emulsions/water clarification (16%) were significant issues as well.

From 1996 through 2011, there were an average of 16.1 TSR's per year related to Compressors – with the notable exception of 2007 when there were 44 TSR's related to Compressors. The vast majority (74%) of these TSR's were for fouling issues. It is interesting to note that for the 290 TSR's related to Compressors, there has been a steady increase in the percentage of TSR's that involve interstage cooler fouling – from roughly 15% in the late

1990's to approximately 40% now. An analysis of which stage of compression shows the most fouling points towards the early to mid-stages of compression, with the later stages of compression having relatively fewer issues.

For the 326 TSR's related to Acid Gas Removal, 82% involved caustic units whereas 18% involved amine units. Despite the fact that amine units represent less than 5% of Acid Gas Removal systems, they appear to require more support as demonstrated by the amount of technical support. 87% of TSR submissions were related to fouling issues, 7% for foaming issues, 3% for corrosion issues and 3% for emulsions. As with Compressors, there was a fairly steady level of requests (20.3 TSR's per year related to Acid Gas Removal) – with the notable exceptions of 2008 and 2011 when there were only 8 and 10 TSR's related to Acid Gas Removal, respectively.

As mentioned previously, the overall trend in Light Ends is one of decreasing support. This may be explained in that Nalco's Light Ends program (Actrene®) is particularly well refined and extensively deployed. However, there were "spikes" in Light Ends support in 2001, 2005 and 2009. It is interesting to note that 2001 and 2009 are years in which the total value of chemicals output suffered a percentage decline (-2.5% in 2001 and -9.8% in 2009) and 2005 was a year that saw many unplanned shutdowns due to hurricanes along the Gulf Coast. During times of decreased throughput, the residence time in Light Ends fractionation increases and this often correlates to increased fouling due to reactive monomers such as butadiene residing in towers for longer periods of time. 41% of TSR submissions were related to depropanizer issues, 25% for debutanizer issues, 19% for deethanizer issues, 9% for depentanizer issues and 5% for condensate stripper issues. Nearly all (95%) TSR's from Light Ends are fouling related, though there are some corrosion issues – namely in the depentanizer – where 15 of the 40 TSR submissions were corrosion related, primarily due to acetic acid corrosion in the depentanizer overhead. Trends in Light Ends are telling: deethanizer and depropanizer support are trending downward, whereas debutanizer support is trending upward.

In conclusion – if the trends hold true – we should expect support requests for both Primary Fractionators and Compressors to rise, with support requests for Light Ends to decline (with the possible exception of debutanizers and during industry downturns). Based on the price of crude oil and naphtha, we can essentially predict when our PrimAct™ program will need more support. This database serves as a rich knowledge repository, as well as a reliable tool to determine market trends and appropriate R&D resource allocation, thereby allowing Nalco to focus on issues of great importance and value to our customers.

Finally, we performed a statistical analysis of 116 representative Light Ends deposits from 2008-2011. The purpose of the statistical analysis was to determine what average levels (and expected ranges) of various elements were. A similar exercise based on 46 representative Acid Gas Removal System deposits from 2008-2011 was also carried out.

Iron (Fe)

80% of light ends deposits contained Iron

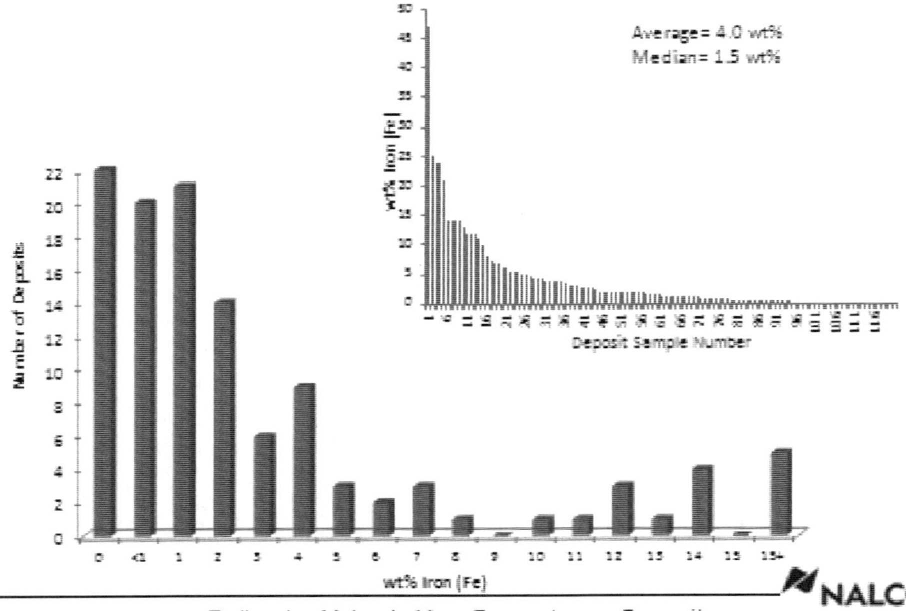

NALCO

Delivering Value in Your Downstream Operation

Nitrogen (N)

23% of light ends deposits contained Nitrogen (>0.3 wt%)
5% of light ends deposits contained Nitrogen > 1 wt%

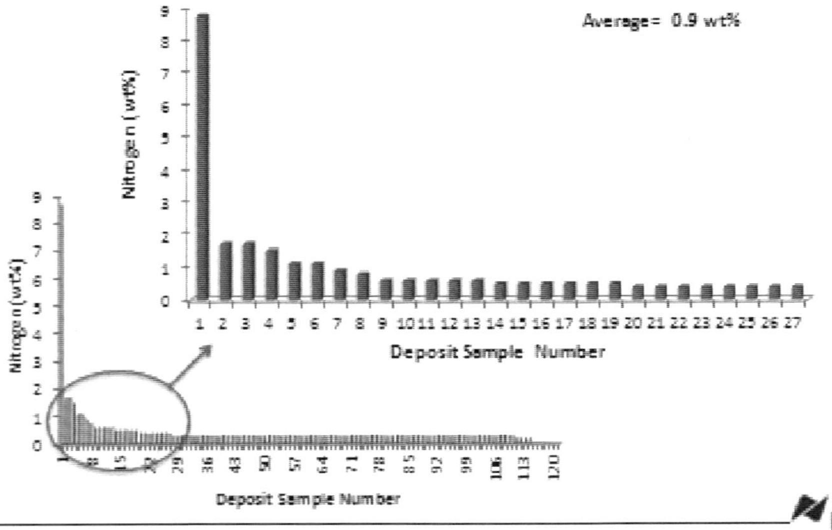

NALCO

Delivering Value in Your Downstream Operation

Sulfur (S)

17% of light ends deposits contained Sulfur

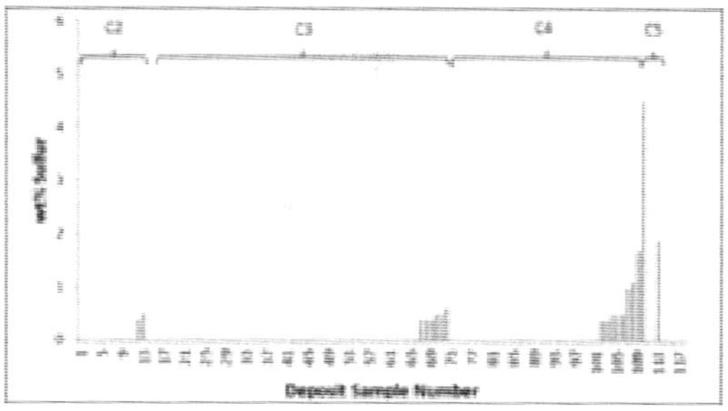

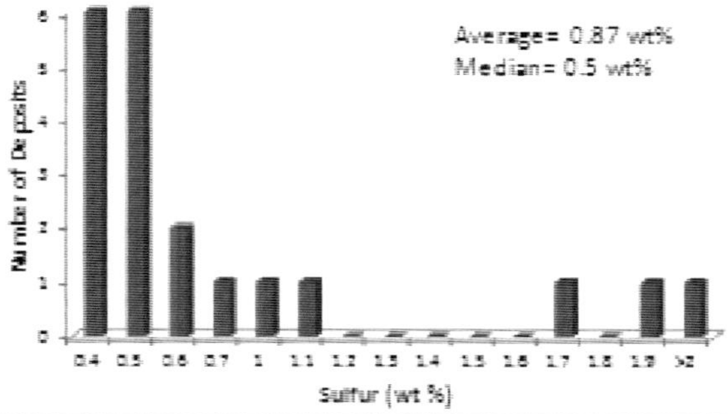

NALCO

Delivering Value in Your Downstream Operation

Silicon (Si)

15% of light ends deposits contained Silicon

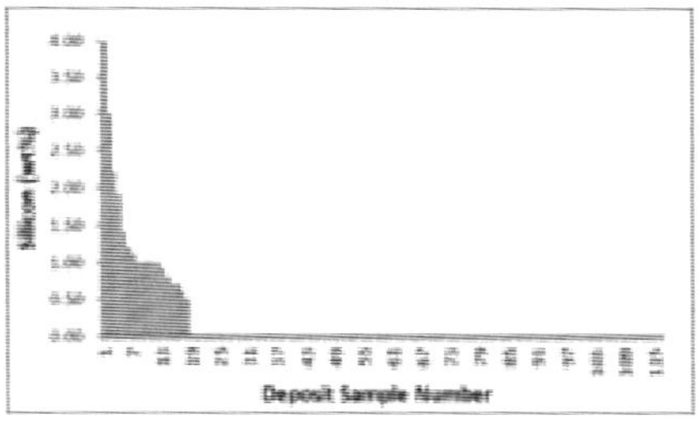

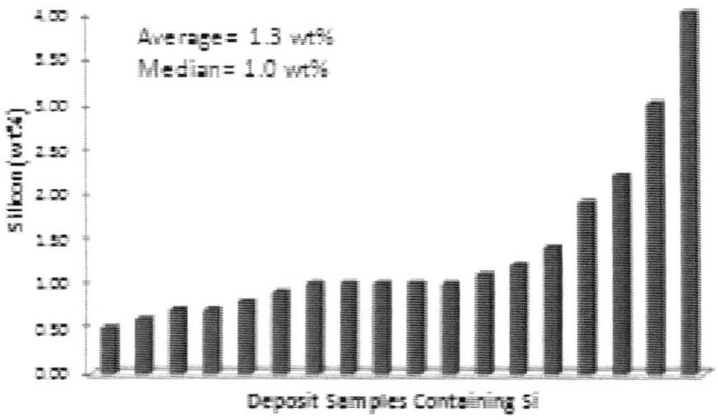

NALCO

Delivering Value in Your Downstream Operation

Additional Elemental Analysis Results

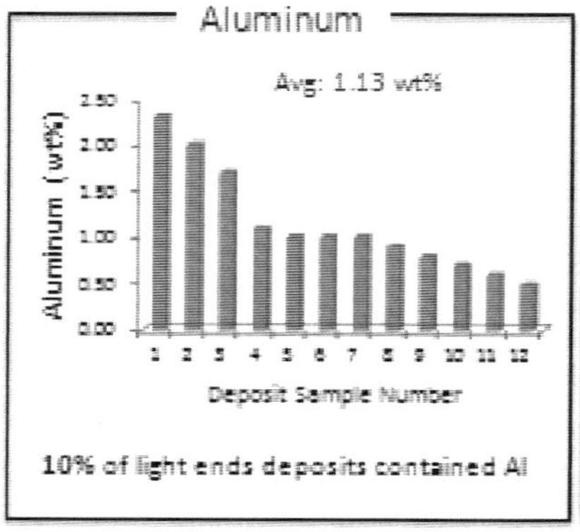

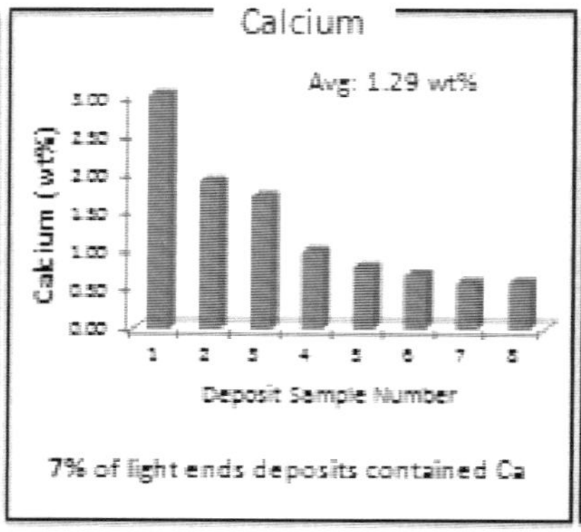

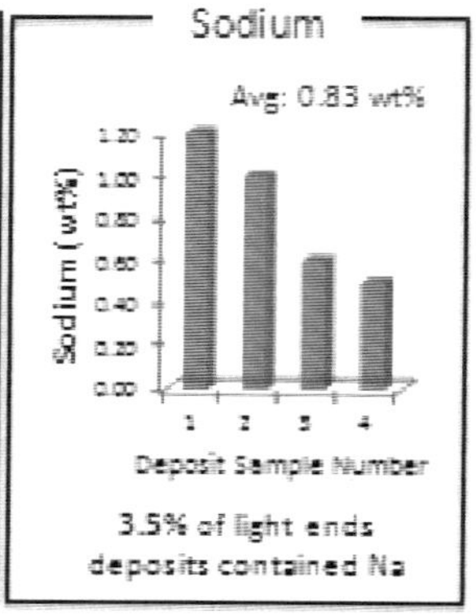

NALCO

Delivering Value in Your Downstream Operation

Oxygen

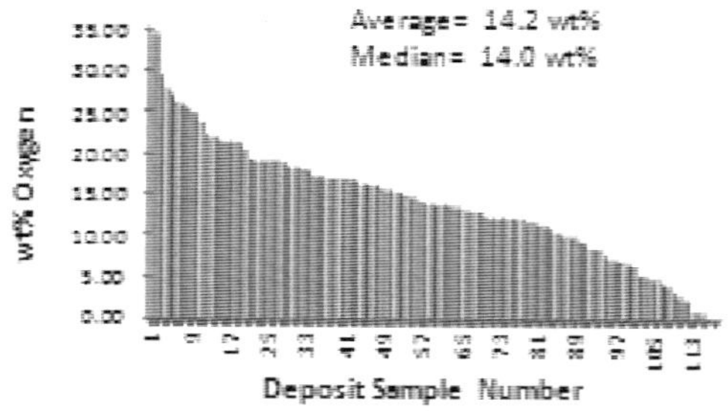

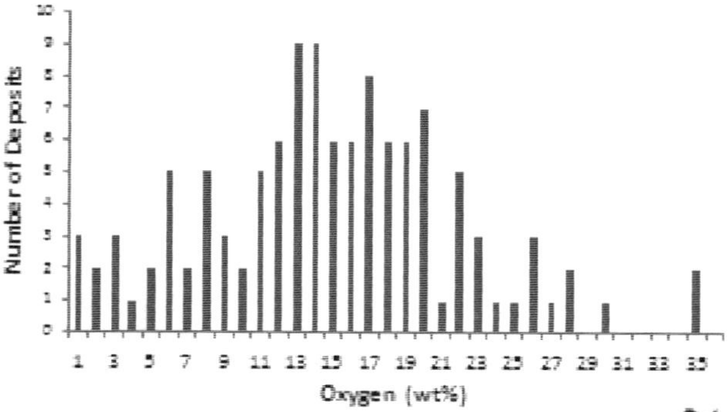

NALCO

Delivering Value in Your Downstream Operation

H/C Ratio

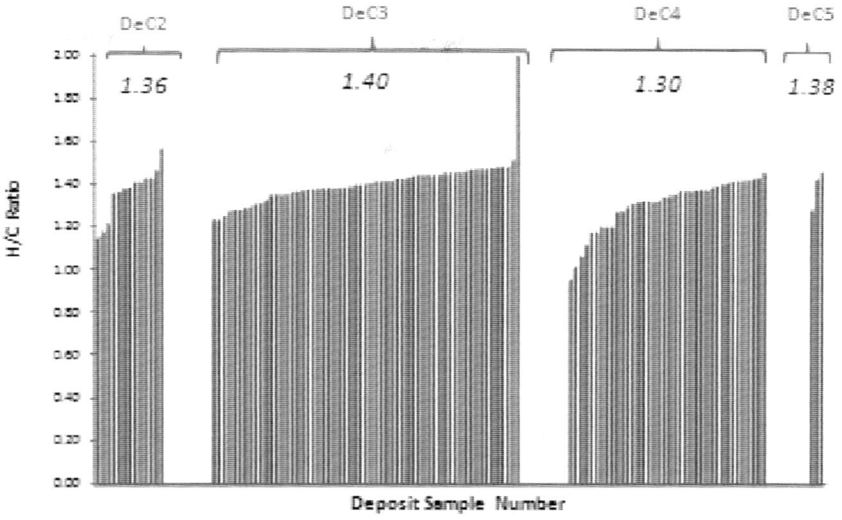

NALCO

Delivering Value in Your Downstream Operation

H/C Ratio

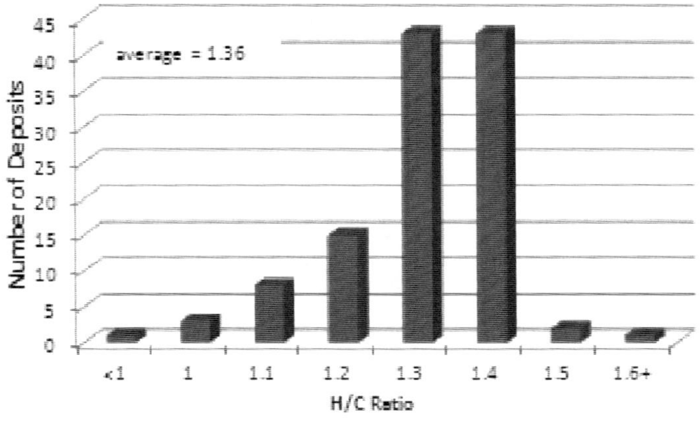

NALCO

Delivering Value in Your Downstream Operation

Loss on Ignition

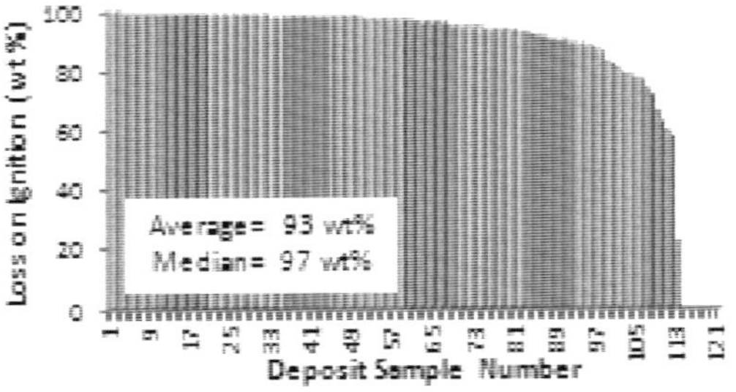

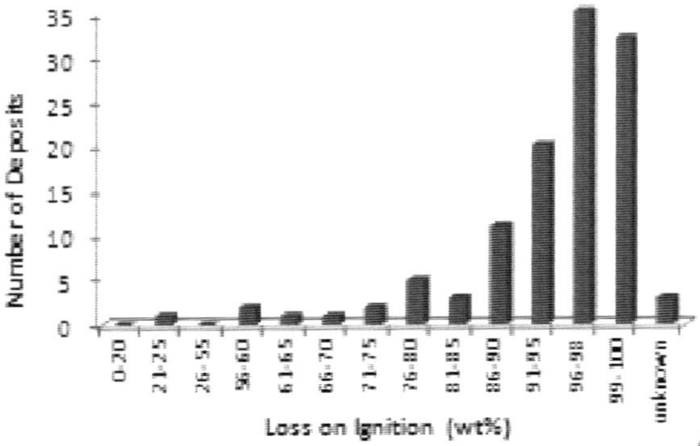

NALCO

Delivering Value in Your Downstream Operation

Dichloromethane Extractable

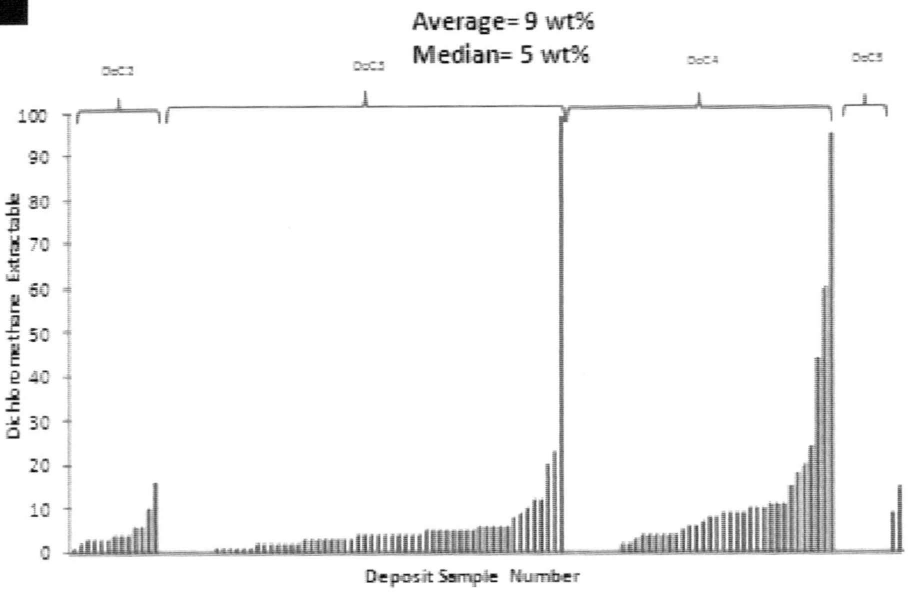

NALCO

Delivering Value in Your Downstream Operation

Acid Gas Removal Deposits

47 Acid Gas Removal Samples Included in Benchmarking Study (2008-present)

- 89% contained Aldol Polymer
- 92% from Caustic Tower vs. Amine units
- Spent Caustic = 40.4%
- Weak Section = 31.9%
- Intermediate = 8.5%
- Strong = 6.4%
- Other = 12.8%

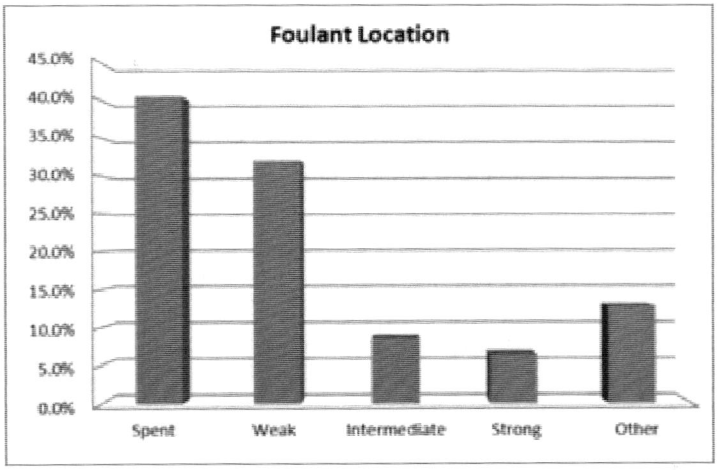

NALCO

Delivering Value in Your Downstream Operation

Average Elemental Composition

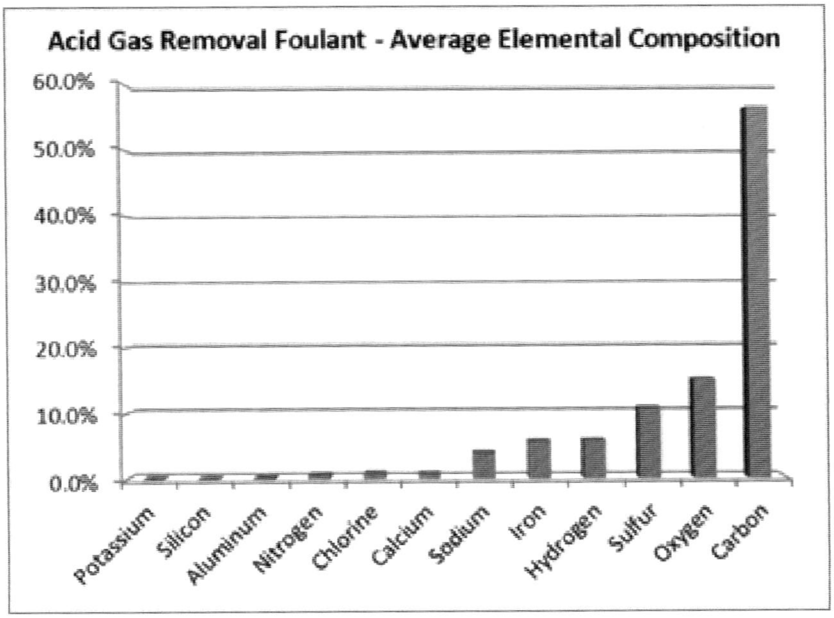

NALCO

Delivering Value in Your Downstream Operation

Sulfur and Iron Content

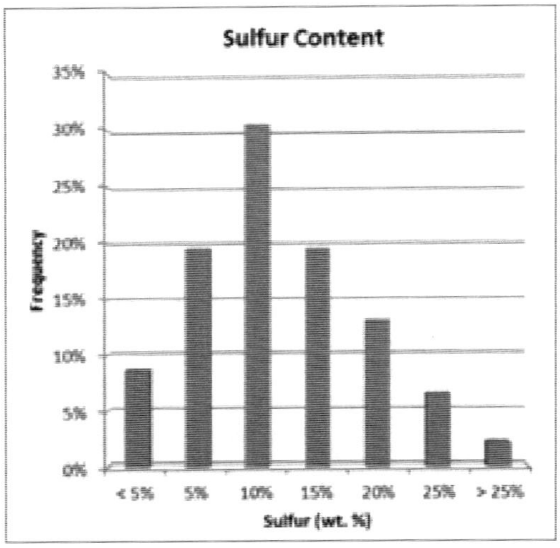

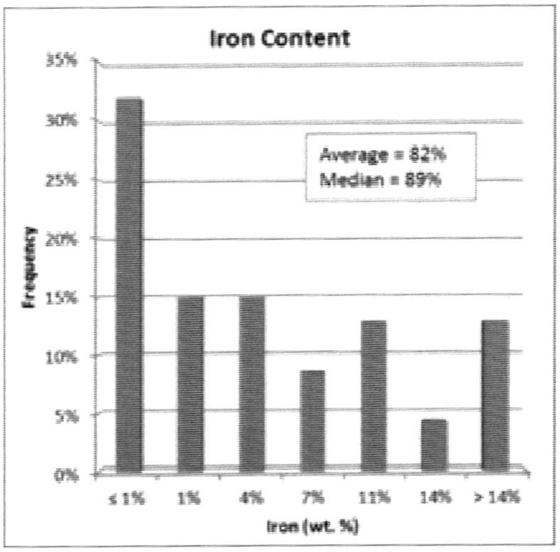

NALCO

Delivering Value in Your Downstream Operation

Oxygen and Nitrogen Content

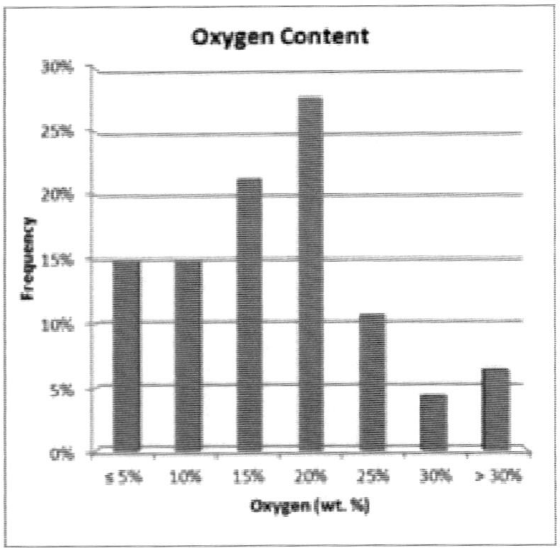

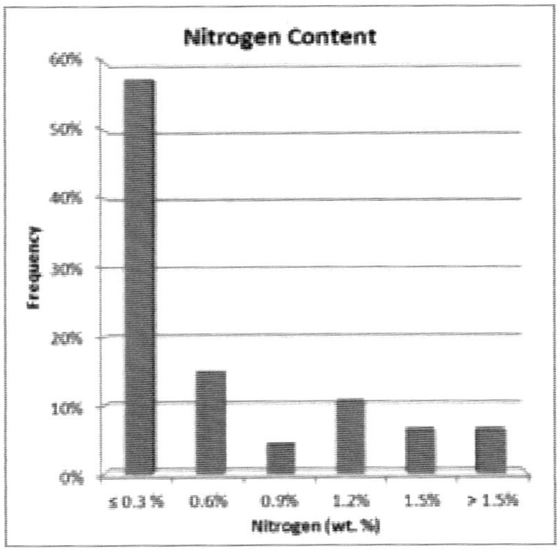

NALCO

Delivering Value in Your Downstream Operation

Solubility and Combustion

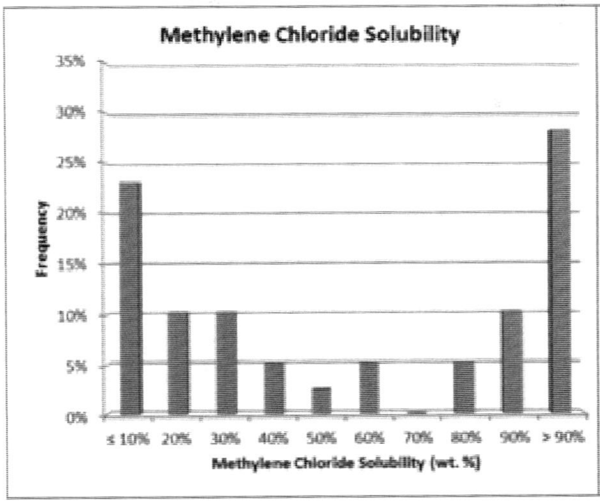

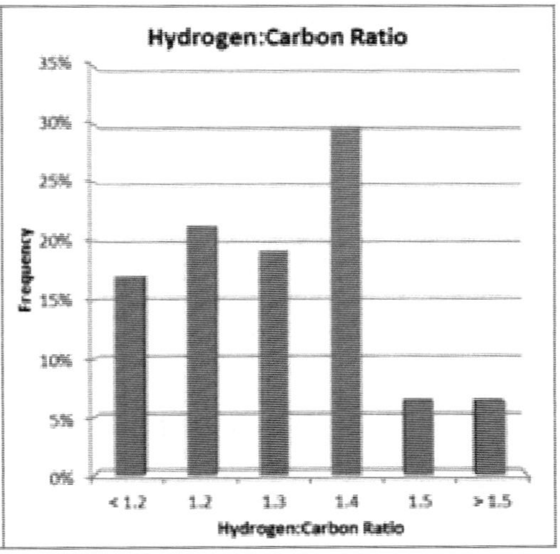

NALCO

Delivering Value in Your Downstream Operation

(26d) An Integrated Approach for Chemical Technology Development

Fuad Mosa, Sabic Technology Center, SABIC, Riyadh 11551, Saudi Arabia

Chemical Technology Development (CTD) projects often span over a decade from initial conception of the idea to commissioning of the commercial plant. To mitigate the high risk involved in developing a new chemical technology from cradle to grave, leading companies have adopted Stage-Gate® System (SGS). However, current as well as future chemical technology development projects are expected to face greater challenges including market uncertainty, less-defined project, better process (product) performance amidst higher technical uncertainties and stricter environmental sustainability requirement. Traditional CTD project management techniques, which are mostly focused on well-defined and predictable projects would not be very efficient in this context. Thus, the need for developing project management techniques that can meet the requirement of next generation CTD projects.

SABIC Technology and Innovation (T& I) Center has recently developed a unique process mapping methodology based on Stage-Gate® System for its next generation CTD projects. The developed process mapping is not a fit-for-all type of Stage-Gate® System used by many companies, but rather customizable to meet the requirement of different technology development projects within chemical and allied industry. The process mapping features innovative concepts, such as rigorous First-Principle Predictive Modeling (FPPM), Design of Experiments (DOE), Quality by Design (QbD), Life Cycle Analysis (LCA) amongst others. The incorporation of these concepts within SABIC SGS is not the only unique aspect, but how these concepts are staged and coupled together to reduce risk in CTD projects while accelerating time-to-market. The newly developed process mapping is currently been used in-house for several of SABIC's CTD projects, and results obtained so far has been astonishing. In this work, some examples of how SABIC's new process mapping could be used to reduced CTD projects' risks while accelerating time-to-market would be demonstrated.

MISAPPROPRIATION OF TRADE SECRETS BY DEPARTING EMPLOYEES OR OTHER AGENTS

Teresa J. Lechner-Fish
Chris P. Perque
Gardere Wynne Sewell, LLP
Houston, TX
www.gardere.com
Telephone 713.276.5500
Email tlechnerfish@gardere.com
Email cperque@gardere.com

I. INTRODUCTION

This paper discusses a common business tort: misappropriation of trade secrets by departing employees or other agents. A trade secret may be any information that can be used to economic advantage. However, to be entitled to trade secret protection, such information must be maintained as a secret. Valuable proprietary information can be lost when employees or other agents depart to a competitor. Accordingly, protection of such proprietary information must begin at the onset of the employee's or agent's employment and continue throughout the employment. Protection and enforcement of trade secrets is discussed for various jurisdictions with an emphasis on recent cases in California and Texas.

II. MISAPPROPRIATION OF TRADE SECRETS BY DEPARTING EMPLOYEES OR OTHER AGENTS

Most states have adopted some form of the Uniform Trade Secret Act (UTSA), however, Massachusetts, New Jersey, New York and Texas have not adopted the Act. A few states including Texas and New York refer to the Restatement of Torts.

Preliminary injunctions may be obtained in special circumstances to prevent irreparable harm to the trade secret owner. Examples of recent preliminary junction cases in California and Texas are discussed below.

A. California

Like most states, California has adopted the Uniform Trade Secrets Act. *See* California Uniform Trade Secrets Act, CAL. CIV. CODE §§ 3426 *et seq*. Under the California Uniform Trade Secrets Act,

> (b) "Misappropriation" means: (1) Acquisition of a trade secret of another by a person who knows or has reason to know that the trade secret was acquired by improper means; or (2) Disclosure or use of a trade secret of another without express or implied consent by a person who: (A) Used improper means to acquire knowledge of the trade secret; or (B) At the time of disclosure or use, knew or had reason to know that his or her knowledge of the trade secret was: (i) Derived from or through a person who had utilized improper means to acquire it; (ii) Acquired under circumstances giving rise to a duty to maintain its secrecy or limit its use; or (iii) Derived from or through a person who owed a

duty to the person seeking relief to maintain its secrecy or limit its use; or (C) Before a material change of his or her position, knew or had reason to know that it was a trade secret and that knowledge of it had been acquired by accident or mistake.

'Trade secret' means information, including a formula, pattern, compilation, program, device, method, technique, or process, that:

(1) Derives independent economic value, actual or potential, from not being generally known to the public or to other persons who can obtain economic value from its disclosure or use; and

(2) Is the subject of efforts that are reasonable under the circumstances to maintain its secrecy.

See § 3426.1(b) and (d); *see also ReadyLink Healthcare v. Cotton*, 24 Cal. Rptr.3d 720, 727-28 (Cal.Ct.App. 2005).

In *ReadyLink*, the plaintiff was a licensed healthcare service provider with a nurse staffing division, which provided healthcare facilities and professionals with nurses. *See ReadyLink*, 24 Cal.Rptr.3d at 723. The plaintiff recruited, interviewed and hired nurses as its own employees and assigned them to its customer healthcare providers who requested nurses. *Id.* The plaintiff invested time, money and effort in developing its database of nurses and healthcare provider customers. *Id.* at 723-24. The database was used to recruit nurses and to service its clientele. *Id.* at 723.

The plaintiff also employed a sales force to recruit nurses, secure new business and service existing customers. *See ReadyLink*, 24 Cal.Rptr.3d at 724. Plaintiff hired the defendant, and, about 6 months later, it moved him into a nurse recruiting agent position. *Id.* The defendant signed 1) upon his employment, an employment agreement and an employee confidentiality, non-disclosure and non-circumvention agreement; and 2) upon his move into the nurse recruiting agent position, an agent agreement and another proprietary and confidential information, non-disclosure and non-circumvention agreement. *Id.*

About 14 months later, the plaintiff fired defendant for stealing company records containing proprietary and confidential information. *See ReadyLink*, 24 Cal.Rptr.3d at 724. While employed with the plaintiff, the defendant was caught on videotape entering company private offices at night and copying payroll practices, confidential documents and notes. *Id.* The defendant authorized a search of his residence and storage locker. *Id.* During the search, the police recovered 1-2 boxes of plaintiff's stolen documents. *Id.* Although the police investigated the theft, the defendant was not criminally prosecuted. *Id.*

Upon termination of his employment, the defendant signed a declaration acknowledging 1) his misappropriation and improper use of plaintiff's proprietary and confidential information, 2) his attempt to form his own company in competition with plaintiff, and 3) his improper solicitation of plaintiff's employees to staff his company and investors to finance his company. *See ReadyLink*, 24 Cal.Rptr.3d at 724. A mere two weeks after his termination, the defendant began working at Registry Network, a competitor, despite agreeing in his declaration to not be employed at any other healthcare staffing company. *Id.* The plaintiff brought suit against defendant and Registry Network for, *inter alia*, violating the California Uniform Trade Secrets Act, alleging defendant stole confidential information regarding the company's finances, clients, employee contacts, payroll practices and business methods.[1] *Id.* at 724-25. The plaintiff requested injunctive relief, as well as damages, prohibiting the defendant from disclosing the plaintiff's proprietary information to Registry Network and from soliciting plaintiff's employees and customers. *Id.* at 725. The trial court granted a preliminary injunction. *Id.*

Upon interlocutory appeal, the defendant contended that the trial court abused its discretion in granting the preliminary injunction. *See ReadyLink*, 24 Cal.Rptr.3d at 722 & 726. The Court of Appeals affirmed the preliminary injunction order, as modified. *Id.* at 723 & 730. The Court stated:

> ...***in deciding whether to issue a preliminary injunction, a trial court weighs two interrelated factors: the likelihood the moving party ultimately will prevail on the merits, and the relative interim harm to the parties from the issuance or nonissuance of the injunction***.

Id. at 726-27 (quoting *Whyte v. Schlage Lock Co.*, 101 Cal.App.4th 1443, 1449-50, 125 Cal.Rptr.2d 277 (Cal.Ct.App. 2002)) (emphasis added). The Court found ample evidence supporting the trial court's finding that plaintiff would likely prevail on its misappropriation of trade secrets claim:

(1) in defendant's declaration upon termination, the defendant admitted that he misappropriated proprietary and confidential information from plaintiff, and that a disclosure of such information violated local, state, and federal law, and he also admitted that, as a consequence of misappropriating such information, it would not be possible for him to engage in the healthcare staffing business without indirectly or inadvertently using or disclosing such information;

(2) in plaintiff's declarations supporting the injunction, plaintiff presented overwhelming evidence that the defendant intended to start his own nurse staffing business in competition with plaintiff using plaintiff's trade secret information, and that he solicited plaintiff's employees and nurses using plaintiff's proprietary and confidential information such as plaintiff's databases containing lists of nurses, employees, and healthcare facility customers, compilations of compensation, employment preferences, contact information, nurse applications and tests, and plaintiff's per diem program;

(3) the supporting declarations sufficiently establish that plaintiff's trade secret information has potential economic value because the plaintiff went to great expense to compile the data, and that the information would enable a competitor to recruit nurses and employees away from the plaintiff; and

(4) the defendant entered into two employment agreements in which he agreed not to use or disclose plaintiff's proprietary and confidential information.

Id. at 728. Therefore, "it is reasonably likely that the plaintiff will prevail on its misappropriation claims." *Id.* at 730.

After finding that the plaintiff would likely prevail on its trade secrets misappropriation claims, the Court reviewed the relative interim harm to the parties from the issuance or nonissuance of the

---

[1] After plaintiff notified Registry Network of the defendant's theft of its confidential and proprietary information, Registry Network terminated the defendant's employment. *Id.* at 725.

injunction, and found ample evidence that defendant posed a threat to plaintiff's business operations at the time of the injunction hearing, including:

(1) the defendant intended to create his own competing business, and he had taken significant steps to carry out his business plan, even though he had agreed not to commit such illicit acts;

(2) shortly before his termination, the defendant was caught on videotape stealing plaintiff's documents, and a search of his home and computer revealed he had misappropriated plaintiff's documentation and information;

(3) the defendant acknowledged he misappropriated proprietary and confidential information, and he solicited plaintiff's employees and nurses using the information even though he had agreed not to use or disclose the information; and

(4) right after leaving plaintiff, the defendant secured a job at a competitor and continued soliciting plaintiff's employees and customers using plaintiff's proprietary information, while simultaneously pursuing his plan of creating his own healthcare staffing business.

Id. at 732. Because the plaintiff also satisfied the irreparable harm requirement, the Court affirmed the preliminary injunction order, as modified. *Id.* at 734.

California companies should guard the secrecy of their proprietary and confidential information by marking confidential documents as "confidential to company" and storing original documents in a vault, limiting access to information on its internal computer system to persons with appropriate system identification and passwords, requiring persons entering the company premises to identify themselves and to wear a badge, checking materials in and out of the premises, and using confidentiality agreements with employees upon employment and, again, upon reassignment to another position.

For particularly sensitive information, companies should use video cameras to record activity in areas where confidential information is kept, or where copiers are located. If an employee is caught stealing such information, the employer should require the employee to sign a declaration admitting to wrongdoing and agreeing to non-compete provisions upon termination.

California employees should avoid accessing their employer's proprietary and confidential information without its express or implied consent to do so.

B. Texas

Texas—like New York—has not adopted the UTSA—it applies common law referring to the Restatement (First) of Torts. To establish misappropriation of trade secrets, a plaintiff must show:

(1) existence of trade secret;

(2) trade secret was acquired through a breach of a confidential relationship or was discovered by improper means; and

(3) defendant used the trade secret without the plaintiff's authorization.

See IAC, Ltd. v. Bell Helicopter Textron, Inc., 160 S.W.3d 191, 197 (Tex.App. –Fort Worth 2005). Texas courts have adopted the definition of a trade secret from the commentary to Section 757 of the Restatement of Torts:

> Any formula, pattern, device or compilation of information which is used in one's business and presents an opportunity to obtain an advantage over competitors who do not know or use it.

See IAC, 160 S.W.3d at 197 (quoting *In re Bass*, 113 S.W.3d 735, 739 (Tex. 2003)); *see also* RESTATEMENT (FIRST) OF TORTS § 757 cmt. b (1939). To determine whether information is entitled to trade secret protection, courts consider the *Bass* factors:

(1) the extent to which the information is known outside the holder's business;

(2) the extent to which it is known by employees and others involved in the holder's business;

(3) the extent of the measures taken by the holder to guard the secrecy of the information;

(4) the value of the information to the holder and its competitors;

(5) the amount of effort or money expended by the holder in developing the information; and

(6) the ease or difficulty with which the information could be properly acquired or duplicated by others.

See IAC, 160 S.W.3d at 197-98 (citing *In re Bass*, 113 S.W.3d at 739); *see also* RESTATEMENT (FIRST) OF TORTS § 757, cmt. b.

In *IAC*, plaintiff was in the helicopter design and manufacture business. *See IAC*, 160 S.W.3d at 196. The defendants were in the helicopter maintenance and service business, and were trying to be in the helicopter replacement blade business. *Id.* In 2001, Jim Van Horn formed Van Horn Aviation, LLC to develop a 206B replacement blade. *Id.* Randy Stevens, principal of IAC, Ltd., agreed to manufacture and market the replacement blade. *Id.* At the time, IAC, Ltd. was a customer service facility (CSF) for the plaintiff. *Id.* As a CSF, IAC, Ltd. received proprietary data relating to the plaintiff's 206B rotor blade. *Id.* Stevens agreed to use the proprietary data disclosed by plaintiff solely for the purpose of repairing plaintiff's helicopters. *Id.*

Knowing that plaintiff's OH-58 blade is similar to its 206B blade, Van Horn purchased a Depot Maintenance Work Report (DMWR) for plaintiff's OH-58 helicopter from Newport Aeronautical Sales. *Id.* Van Horn used the OH-58/206B contour and twist data and 206B sealants and adhesives for the replacement blade. *Id.* at 197. When the plaintiff learned that defendants were making a replacement blade, it sued for, *inter alia*, misappropriation of trade secrets, and theft of trade secrets. *Id.* The trial court granted a temporary restraining order, and, about two months later, a temporary injunction. *Id.*

Upon interlocutory appeal, the defends contended that, *inter alia*, the trial court erred in granting the temporary injunction. *See IAC*, 160 S.W.3d at 197. The Texas Court of Appeals affirmed the temporary injunction. *Id.* at 203. The Court of Appeals stated:

To be entitled to a temporary injunction, the applicant must plead a cause of action and show a probable right to recover on that cause of action and a probable, imminent, and irreparable injury in the interim...

Id. at 197 (internal citations omitted) (emphasis added). The Court found that the plaintiff's evidence, when viewed in light of the *Bass* factors, supported the trial court's finding that the 206B data that plaintiff disclosed to IAC, Ltd. and the DMWR obtained by Van Horn are entitled to trade secret protection at least until a trial on the merits. *Id.* at 198. With respect to the 206B data, the plaintiff produced evidence showing the measures it takes to guard the secrecy of the data, the amount of effort plaintiff expended in developing the data, the data's value, and the difficulty with which the data could be duplicated by others:

(1) plaintiff guards the secrecy of its data by storing the originals of its drawings and specifications in a vault, posting security guards at its plants, requiring persons entering the plant to identify themselves and wear identification badges, checking material going in and out of the plant, limiting access to data on its internal computer system to persons with appropriate system identification and passwords, and using confidentiality agreements with its vendors to whom it supplies proprietary data;

(2) one of plaintiff's engineers testified that thirty to forty thousand hours were required to develop the design of the blade, and another thirty to forty thousand hours were needed to develop techniques for its production;

(3) one of plaintiff's sales and marketing employees estimated that the replacement blade market was worth approximately twelve million dollars a year; and

(4) expert testimony showed that it would not be possible to duplicate the blade specifications with the necessary degree of precision through reverse engineering.

Id. at 198. With respect to the DMWR, the plaintiff produced evidence of the measures it takes to protect the secrecy of the data contained in the DMWR and the extent to which the information is known outside of plaintiff's business:

(1) plaintiff showed that the DMWR was created using plaintiff's drawings that would have been marked "confidential to Bell" and that the DMWR stated on its face that it was not available through Attorney General publication centers; and

(2) plaintiff's evidence also revealed that the drawings used to create the DMWR were not available through a Freedom of Information Act (FOIA) request, and that Newport Aeronautical, Van Horn's alleged source of the DMWR, provided the DMWR only on a "need to know basis" and did not provide electronic access to it.

Id.

The Court also found that the plaintiff's evidence supported the trial court's finding that the defendant acquired the 206B data and DMWR through a breach of a confidential relationship or discovered it by improper means. *See IAC*, 160 S.W.3d at 199. With respect to acquiring the information through a breach of a confidential relationship, the plaintiff produced evidence of IAC, Ltd.'s probable breach of the confidential relationship with the plaintiff to obtain the 206B data:

(1) several witnesses testified, and plaintiff's Customer Service Facility Agreement showed, that plaintiff disclosed 206B data to IAC, Ltd. so that IAC, Ltd. could repair plaintiff's helicopters;

(2) IAC, Ltd. had an obligation to keep the plaintiff's data confidential pursuant to a confidentiality agreement;

(3) Stevens, the president of IAC, Ltd., was a partner with Van Horn in the defendant's 206B replacement blade project;

(4) IAC, Ltd. kept the 206B data in its library, and Van Horn had access to the data; and

(5) the contour and twist data for the 206B blade is identical to that for the replacement blade.

Id. With respect to acquiring the information through improper means, the plaintiff also produced evidence of Van Horn's likely improper means to obtain the DMWR:

(1) the DMWR was not available through Attorney General publication centers;

(2) the drawings upon which the DMWR was based were not available through a FOIA request; and

(3) Newport Aeronautics provided the DMWR only on a "need to know" basis.

Id.

In addition, the plaintiff produced some evidence that defendant used the 206B data and DMWR without the plaintiff's authorization:

(1) the defendants used contour and twist data contained in both the 206B data disclosed to IAC, Ltd. and the DMWR and that they used primer and adhesive specifications present in the 206B data;

(2) Van Horn created his certification plan to submit to the Federal Aviation Administration for the 206B project on the same day he purchased the DMWR;

(3) IAC, Ltd.'s was expressly prohibited from using the 206B data pursuant to a confidentiality agreement, and Van Horn was not authorized to use the data in the DMWR.

Id. at 199-200. "Because [plaintiff] presented some evidence to support each element of its trade secret misappropriation claim, it has shown a probable right to recover under this cause of action." *Id.* at 200.

After finding that the plaintiff would probably recover on its trade secrets misappropriation claim, the Court reviewed whether the plaintiff met its burden of showing that a probable, imminent, and irreparable injury would result before trial in the absence of a temporary injunction. *See IAC*, 160 S.W.3d at 200. Plaintiff's evidence showed that the defendants had possession of plaintiff's data entitled to trade secret protection and they were actively using that information to compete with plaintiff in the replacement blade market. *Id.* Further, plaintiff's evidence also showed that defendant's replacement blades were going to be available for shipment around June 15, 2004 and would be priced lower than plaintiff's blade:

> [Plaintiff's] Executive Director of Integrated Product Solutions testified that the creation of a lower-priced 206B replacement blade market would undermine the goodwill that Bell has with its clients, leading to an incalculable loss of business goodwill.

Id. at 200. Because the plaintiff also satisfied the imminent and irreparable injury requirement, the Court held that the trial court did not abuse its discretion by granting the temporary injunction. *Id.*

Texas companies should guard the secrecy of their proprietary and confidential information by marking confidential documents as "confidential to company" and storing original documents in a vault, limiting access to information on its internal computer system to persons with appropriate system identification and passwords, requiring persons entering the company premises to identify themselves and to wear a badge, checking materials in and out of the premises, and using confidentiality agreements with employees and vendors. For particularly sensitive information, companies should use video cameras to record activity in areas where confidential information is kept, or where copiers are located.

Texas vendors, that are subject to a confidential agreement, should guard the secrecy of the other party's confidential information as discussed above.

Chris Perque is a partner with Gardere Wynne Sewell LLP in Houston, Texas specializing in intellectual property and commercial litigation.

Teresa Lechner-Fish is an associate with Gardere Wynne Sewell LLP in Houston, Texas specializing in intellectual property and commercial litigation.

The views expressed herein are solely those of the authors and do not necessarily reflect the views of Gardere Wynne Sewell LLP or its clients.

(36b) Environmental Liability: Regulation, Compliance and Litigation Risk

Mary Ellen Ternes, McAfee & Taft, Oklahoma City, OK

Compliance with environmental regulations, while necessary and helpful in minimizing litigation risk, does not completely insulate industry from litigation. This presentation will review potential pathways towards unwanted administrative and judicial entanglements regarding which even the most compliant companies should be aware, including enforcement disputes, citizen suit actions and common law claims.

Title: Propellant Systems for Household Spray Products
Submitter's E-mail Address: cjabraham1@aol.com
Preferred Presentation Format: Oral
Keywords: Hazards, Standards and Total Release Foggers
Session Selection: Litigation Perspectives for Chemical Engineers
AICHE – Houston, TX April 1-5- 2012

PROPELLANT SYSTEMS FOR HOUSEHOLD SPRAY PRODUCTS

C. J. Abraham, Scientific Advisory Services, Ltd, Great Neck, NY

Abstract

The problem delineated in this paper involves consumers throughout the United States who use and are exposed to bug bombs or bug foggers (also known as total release foggers orTRF's). They are the most dangerous aerosol products on the market. We will focus on the defective design of the formulation, the state of technology available to eliminate the risks, the defects in the testing of the product, the false information that is submitted to the EPA and the lies stated in the application process and registration of pesticide products.

The most popular propellant system for total release foggers for indoor use consists of propane, butane and isobutene. Even though the propellent system consists of up to 35% of the ingredients, applications to the EPA for approval of insecticides designate the propellant system as inert. This designation is incorrect and fraudulent. In addition to the products being defectively designed, they contain a hidden hazard resulting in fires, explosions and traumatic burn injuries.

There are several resolutions which can make the product either completely safe or minimize the risk of consumers exposing themselves to explosions and toxic vapors. This involves alternative safer propellant systems and significant changes in the testing of the product, proper reporting of information in the MSDS's, proper testing and significant changes in the warnings.

Introduction

Over the last forty years this author has been involved with dozens of fires and explosions involving a popular propellant used in a number of products including total release foggers also known as TRF's. The propellant is composed of propane, butane and isobutene. The insecticide is purchase by the lower income segment of our population.

The last case that I was involved in was typical of what could happen with the TRF's. Two women, aged 25, and 26, were severely burned when the explosive and extremely flammable propellant from a popular insecticide product came in contact with an ignition source in the home of the mother of one of the women about seven hours after her mother released a number of canisters in her home. The women were burned over 70% to 90% of their bodies. Both women sued the manufacturer, retailer and distributor of the TRF insecticide.

Both women lived in a rural area of, Nebraska and the mother lived in Iowa. Prior to arriving at the mother's home, they all met at a store in between. The women were not informed of the canisters,. The women proceeded to the mother's residence to pick up two dogs.

Once the women arrived at the residence, they went in through the front door. The house needed cleaning up so both women decided to surprise the mother with a clean house. when she arrived sometime later.

Prior to the explosion, neither women detected any odor in the house other than the odor from the animals. After some time, both of them became exhausted and lay down on the couch. They woke up to the barking of dogs. They rose up to see what the dogs were barking about. When they were approximately three to four feet from the rear of the house, they heard the furnace "click" on. An explosion and flashover followed and both of them experienced the results and extreme heat of the flashover. As soon as they heard the explosion, one of them noticed that she was on fire. They went out the back door dropped and rolled on the ground.

It is important to note that neither of the women detected any odor of a liquid, chemical or vapor. That includes the gasoline which was located in a plastic container placed on the dryer in the washroom. There was also a very small amount of the gasoline remaining on the floor in the bathroom from a prior spill. Neither women used the hot water or adjust the thermostat. In addition, the lights in the living room, kitchen and bathroom were on when they arrived.

The mother never told her daughter or her freind that she set-off eight (8)TRF's throughout the house prior to her leaving. The mother's grandchildren were expecting to visit her the following week and they were allergic to spiders. The mother had purchased the TRF's at a "Dollar" store. The instructions suggest that the consumer can use one fogger per room. The label states that the four-can package is able to treat 2,000 cubic feet of unobstructed area. The subject house is approximately 1,200 square feet. There are no statements indicating what would happen if extra TRF's were used throughout the house.

The explosion and fire that occurred destroyed the home. Initially, the fire department did not perform a complete analysis with reference to the cause and origin. For example, if gasoline vapor was the cause of the fire, why wasn't the container of gasoline on top of the dryer involved with the fire? The fact is the gasoline container was still intact after the house was destroyed. The Fire Department concluded that the cause of the fire was the propellant from the TRF coming in contact with the ignition source of the furnace.

Background- The Product

Total release foggers can be purchased in many grocery, hardware stores and chain stores They are relatively inexpensive and have been used for several decades in lieu of using a professional exterminator. In order to properly use those products, one must disconnect all the electrical appliances. This includes refrigerators, freezers, lights and anything else that uses electricity. In addition, the furnace, gas stoves, hot-water heaters and all other sources of ignition must be turned off. It should be noted that all of the above is the responsibility of laypeople with limited or no education in this area. The total release foggers are turned on throughout the apartment or house and the area is left completely closed. The owners return a number of hours later with the expectation that the cockroaches, spiders and other insects are killed. The manufacturer's instructions state and assume that after a number of hours the previous explosive environment is safe to enter and that everything that was turned off previously can now be turned on again. Not one manufacturer informs the consumer that an explosion-proof ventilation system should be used to make sure that most, if not all, of the concentration of the explosive propellant is dissipated. The manufacturers also know that the average layperson does not have any type of gas-detection equipment, so he would not be equipped to even measure or detect explosive gases. The gases are invisible and are not odorized. Therefore, the consumer is exposed to a hidden danger that is known to the manufactures but not known to the consumers.

The Consumer Product Safety Commission (CPSC) has reported accidents involving similar propellant systems from the early 70s to the present time. One of the many cases that the undersigned has consulted on was reported in the United States District Court Decision: Allison Kowak, et al Plaintiffs vs. Fabrege U.S.A. and Precision Valve Corp., Defendants. Civ. No. CV 90-1919. Nov. 13, 1992. In addition, a demonstration was performed by the undersigned on CBS TV shortly after the appellate decision was published (812F.Supp. .942).

For many decades, thousands of explosions have occurred where total release foggers (TRF's) were used in apartments and homes throughout the United States. These explosions resulted in fires, destruction of property, burn injuries and death. They occurred mostly in apartments of the segment of our population that is made up of lower-income families that cannot afford professional exterminators.

In most instances, the TRF is composed of approximately 9% mineral spirits and 0.515% cypermethrin and a 35% combination of propane, butane and isobutane. The sole purpose of the propellant is to spread out and distribute the active ingredient, cypermethrin. In essence, the inert ingredients of the product account for 99.485% and the active ingredient accounts for 0.515%.

In addition to the above, major manufacturers, in every one of their applications to the EPA for approval purposes, indicated that their products incorporated only the active ingredients and not the propellants as the ingredients to be evaluated. In fact, the applications submitted stated that the inert ingredients, mainly propellants, amounted to 99.3% or more.

From a formulation standpoint, everything that is in the container for any specific use is the total amount of each ingredient as retailed to the public. For example, in all the applications of a TRF by more than one major manufacturer, propane, butane and isobutane should have been incorporated as part of the application process to the EPA and not designated as an inert ingredient. If it was an inert ingredient the thousands of explosions, fires and resulting injuries over the last 30+ years would never have occurred.

The Propellant

Manufacturers, in most instances, fail to place the EPA and the consumer on notice that the highly flammable and explosive gaseous propellants are invisible and much heavier than air. They also travel along the lower part of the floor unless disturbed. This creates a hidden danger to the consumers but is well known to the manufacturers. More importantly, those facts have been concealed from the EPA in the application and approval process.

What makes it more difficult to detect is the fact that the explosive formulation is invisible and not odorized. In addition, the vapor density of the propellant and active ingredients are heavier than air and tend to fall to the lowest part of a room or floor. In many instances, the vapor can actually flow down a stairway and wind up at the lowest part of the floor below which is usually the furnace. Contrary to the manufacturer's inferences, the TRF is not distributed evenly throughout a room from floor to ceiling. In addition, the system does not follow the ideal gas laws.

The flashpoint for the subject propellant, as stated on the Material Safety Data Sheet, is (minus) -140ºF. The flashpoint of a volatile liquid is the lowest temperature at which it can vaporize to form an ignitable mixture in air. Thus, the propellant can ignite at temperatures much lower than normal room temperatures.

Propane has a molecular weight of 44.10 grams per mol as opposed to 28.97 for air so it is about 1.5 times heavier than air. Iso-butane and butane have a molecular weight of 58.12 grams per mole so it is about 2 times heavier than air. When released in air, the propane / isobutene propellant has a negative buoyancy and thus falls under the effect of gravity.

In his textbook "Evaluation of the Effects and Consequences of Major Accidents in Industrial Plants" (Publisher: Elsevier Science, 2008, Chapt. 6, pp. 195-248), Joaquim Casal discusses the flow of heavier than air gasses including propane. The difference in the molecular weight and the negative buoyancy and gravity cause the propellant to fall to the ground level and displace the air. The significant difference in the molecular weight does not allow for the homogeneous mixing of the propellant gas with air.

Aerosols released from a pressurized can will first move in the direction of release due to the momentum of the gas. (Momentum is a quantity in mechanical engineering expressing the motion of a body or system, equal to the product of the mass of a body and its velocity. If the direction of release is pointing upward, the gas will first move up until it loses its momentum under the effects of gravity. At that point, "if the material released is heavier than air, it will first descend to the ground and then spread radially" [Casal page 219].

If the heavier-than-air gas is released in an indoor space, the edges of the cloud will eventually reach the perimeter of the space and then conform to the shape of the bottom section of the space. As Casal states in his book, the mathematical models which predict the actual shape of the cloud as a function of time are complex and require a computer to solve. Most of these models do not account for obstructions in the room. Complications are added if there is any air flow in the room (from a fan, air conditioner, or open window) and if there is a temperature variation in the room from floor to ceiling (because hot air rises). What is known however, is that regardless of the exact shape of the edges of the cloud as time progresses, after several seconds of dispersion, the heavier-than-air gas will displace the air at the bottom of the room and fill the space.

The TRF industry relies upon J. D. Commerford's article entitled: "The Measurement of Propellant Obtained from the Discharge of Total Release Aerosols" published in the July, 1992 issue of the trade magazine "Spray Technology & Marketing." This article could not have been peer-reviewed because the scientific method used was incorrect and inaccurate. The author assumes that the TRFs follow the ideal gas law and the product is distributed throughout each room evenly from floor to ceiling. He also assumes that normal ventilation within all of the environments will dissipate the concentration explosive gases to render them harmless.

Dry air is treated as an ideal gas for scientific analysis even though it is made up of nitrogen, oxygen, carbon dioxide and several other trace gasses. Much of this is done to simplify the analysis and there are many chemical processes that rely on extracting the different gasses from air. Thus, in no case can mixtures of dry air and other gasses be considered to be a uniform ideal gas. As Casal states in his book and illustrates using measurement data from several sources, heavier-than-air gasses have a negative buoyancy in air. When released into the air, heavier-than-air gasses tend to fall under the influence of gravity. The uniform dispersion of the heavier-than-air gasses in an enclosed room cannot occur even after the long periods of time reported.

The heavier-than-air gasses will conform to the floor of the space, displace the air and the volume occupied will be equal to the volume of the heavier-than-air gasses. If the room has a large cross sectional area, the heavier-than-air gasses will occupy a height extremely close to the floor of the room. My expectation when I review the test data reported in the article is that the measurement equipment cannot measure the gas concentrations this close to the floor.

While *not completely incorrect*, the article is certainly misleading. For example, the author of the article says that, "the volatile propellants rapidly flash off leaving finely divided particles of concentrate" [page 29] and "the propellant flashes off into the enclosed air space" [page 28]. The use of the non scientific term "flash off" seems to indicate that the propellant somehow disappears from the room. This is misleading in that the propellant disperses but in no way disappears from the room until a window or door is opened or it seeps through some other opening.

The author of the article also states that there are three schools of thought about where the propellant goes, either falling to the floor, layering itself or being distributed uniformly distributed [page 28]. Again, this is very misleading since it implies that there is disagreement about what happens when a gas is released into an enclosed space. The real situation is that any of these occurrences are possible depending on the properties of the gas.

The author of the article also states, "Instantaneously the volume concentration of propellant in the air are not uniform, but in time the propellant is distributed to all parts of the room." [page 29] This is certainly true but he does not state the obvious implication of this, namely ***that the concentration of the propellant at a certain instant of time and at a particular location of the room can easily be between the lower explosive limit (LEL) and the upper explosive limit (UEL). At this point in space and at this instant of time, if the propellant comes in contact with an ignition source, IGNITION WILL OCCUR.***

There is also a problem with Cummerford's measurement technique in the distance from the propellant source. The sample points were positioned several feet from the source of the propellant. You can model what the exact shape of the plume is as the propellant leaves the can but for the most part, it will not be as you expect. The propellant will go upward because of its momentum as it leaves the nozzle of the pressurized can. Since the propellant is heavier than air, at some height gravity will overcome the momentum and the propellant will begin to fall to the floor. To measure the concentration of propellant where it may be between the lower explosive limit (LEL) and the upper explosive limit (UEL), you would want to put the sample locations within this plume. By placing the sample points several feet from the discharge site, the plume has a large cross sectional area and the concentrations of propellant will clearly be low.

There is a review an article entitled, "Propellant Distribution from Total Release Aerosols Discharged into an Enclosed Chamber, by Richard G. LeVasseur and J. D. Commerford which appeared in the April, 1997 issue of the trade magazine, "Spray Technology & Marketing". This is an update of the previous article for a different propellant and using a different Foxboro instrument for measurement. This article has the same problems as the previous one, namely that the measurement starts too late (although earlier than the previous article) and at a distance that will obviously yield low concentrations.

Lastly, Cummerford did not attempt to test the area with an ignition source because it would have been obvious to the undersigned that there would have been an explosion resulting in injuries to him and others. If that occurred, he would not have been able to write the article for the trade magazine.

It should be noted that the management of the defendant manufacturer used that information for their own product and was testified to be their technical director and research and development manager employed by the defendants. As stated previously,, the assumption that their product follows the ideal gas law is totally incorrect. This can also be demonstrated by taking, for example, a GLADES air freshener and expel some it into a room. The observation of the flow of the ingredients and the vapor follows Casals analysis in the referenced publication.

Chemistry Review of the TRF

A chemistry review of the documents of the manufacturer's reports that the highly flammable explosive repellent under pressure is "Combustible." The manufacturers also state that "exposure to temperatures above 130°F may cause bursting." That is a common statement used on all products containing the propellant. The inference is that the product is not extremely flammable and/or explosive. The technical review branch of the EPA in 2006, reported that the label does not need the signal word: combustible.

The EPA relies on the expertise of the manufacturer to present a complete, accurate and true description of the chemistry, formulation and summary of the testing by the manufacturer.

ANSI Z129.1 1982 Standard titled For Hazardous Industrial Chemicals – Precautionary Labeling states that, "Each hazardous chemical must be considered individually in the light of its own characteristics and customary or reasonably foreseeable handling or use. In such cases, suitable statements should be developed to fit the situation." This is the basic standard of care that the major manufacturers failed to follow.

Since there are millions of organic chemicals and formulations, there is no government agency or international agency that is capable of evaluating each and every formulation. They rely completely on the alleged integrity and honesty of the presenter of the application and/or the manufacturer. In every case, the EPA relies on the manufacturer's integrity of their application in the approval process. They do not have the wherewithal or the facilities nor the time to test the formulations of thousands of applications. In fact, there is no government agency or international agency that is capable of evaluating each and every formulation submitted for approval.

A major corporation's research and development laboratory performed a "Container Stability Determination, Real Time and Accelerated Stability" in their EPA application. The tests also used the "Product Properties Guidelines, Group B-Physical/Chemical Properties." Even though the corporation defined it as a pressurized liquid, not once did they identify the propellant as an extremely flammable/explosive combination of gases when used in a foreseeable protocol. The manufacturers did not properly test the subject product and presented false information in their report.

In one of the cases the undersigned was involved in, the manufacturer reported that the "flammability" is "0" under the section "Hazardous Materials Identification System." The propellant was defined as a "hydrocarbon propellant" with CAS Numbers 75-28-5 and 106-97-8. These gases are a combination of a highly explosive and extremely flammable gas. In section 4 titled "Fire and Explosive Hazards Data," the defendant manufacturer designates that the- "Flammable Limits" is" N/A". One can conclude that it is highly probable that these statements in the MSDS is both fraudulent and false.

More than one manufacturer states that an odor neutralizer is used. This ingredient must also be listed as a component as part of the application submitted to the EPA. The undersigned could not find one document listing the odor neutralizer ingredient.

In addition, manufacturers' MSDS for their TRF products which are described as a total release indoor aerosol insecticide, is formulated with more than 90% of isobutane, butane and propane and is housed within a pressurized container. Under Section III, the MSDS indicates that the vapor density is greater than 1, when in actuality, it varies from 2 to 4. The defendant manufacturer also stateed under "Flammable Limits," that the product is not flammable. That communication to anyone using their products is totally false, misleading and fraudulent.

The EPA's response to all of the misleading information resulted in a suggested hazard statement by them on the label reading, "Contents under pressure. Do not store near heat or open flame. Do not puncture or incinerate container. Exposure to temperatures above 130ºF may cause bursting." There was no mention of the propellants being extremely flammable and explosive because the manufacturer never placed the EPA on notice of that fact and the consumers, with that limited information, released the insecticide throughout the areas of their residence.

Other documents submitted to the EPA by the manufacturer stated that the active ingredients were tetramethrin and cypermethrin. The total of the inert ingredients as listed were 99.30%. The inert ingredients were not defined or identified. However, the manufacturer knew that when their product was released into a room, the "inert ingredients" combined with the active ingredients and distributed the active ingredients within a specified area. In order to obtain the true toxicity of the product, the pressurized liquid concentrate should have been identified. As such, and as elaborated in the "Executive Summary" submitted to the EPA, the information would be questionable. There were many other areas in the EPA application where the flammable characteristics of the repellent, once released, were not properly reported and listed by the manufacturer as not applicable (N/A).

The EPA

The EPA has continually ignored their own experts,, as well as fire officials who are concerned about the potential dangers of explosions and/or fires caused by routine usage or storage of consumer aerosol products. Instead, the EPA responds to pressure from the manufacturers of these products who are resistant to changing warning labels. Extrapolating from data the New York City Fire Department collected concerning the increase in fires and explosions caused by aerosols, an internal EPA memo estimated an excess of 500 fires and/or explosions annually in this country."

In the 1970's, scientists discovered that the chlorofluorocarbons (CFCs) used as propellants in aerosol cans (among other things) were a leading factor in the depletion of the ozone layer. As a result, hydrocarbon propellants (i.e. propane, butane, isobutane) and ether propellants replaced CFCs in aerosol products. The down side to this alternative is that hydrocarbons are extremely flammable substances, and it appears that the public has not been sufficiently warned and educated about their safe use. The

Environmental Protection Agency (EPA) has significant evidence that the propellants can pose significant hazards in normal household use,and yet they have not revealed this information to the public. It is up to the EPA to respond to consumer protection needs. Unfortunately it appears that it reacting to pressure from aerosol manufacturers. The EPA should resist this pressure and change warning labels on many aerosol products, such as total release foggers and space-spray pesticides to warn consumers of the potential dangers.

In response to this danger and an increase in the number of accidents caused by aerosol products, in 1987 the EPA proposed new regulations to improve warning labels on them. But their proposed changes were opposed by the aerosol industry. In an EPA memo one official states, "Industry, through the CSMA [Chemical Specialties Manufactures Association], indicated that they would initiate a lawsuit against PR 87-8 [the EPA's recommended labeling requirements] if it were not rescinded." After much debate, the government succumbed to the legal threats and withdrew their proposal.

The business lobby has been able to cut off all warnings that could aid in decreasing the number of fires and explosions caused by aerosol products. An internal EPA memo states:

"Incident data from the CPSC [Consumer Product Safety Commission] indicates that 60% of pesticide aerosol incidents resulted in explosions that caused injuries requiring medical attention. . . many incidents have resulted due to abuse of the aerosol container.However, if the aerosol cans were obviously marked as 'Flammable' or 'Extremely Flammable' some of the abuse might be prevented. It should be noted that the "abuse" described is foreseeable and known to the manufacturers.

Aerosols using the highly flammable/explosive propellant are used daily in bathrooms, kitchens, and garages, In most cases, the user has no knowledge concerning the propellants in the spray can. According to an article published in Aerosol Age magazine, "Well over half of all aerosols are marked 'FLAMMABLE' or 'EXTREMELY FLAMMABLE'. . .". The article goes on to state, "90% [of aerosol products] contain flammable propellants and over two out of three contain flammable or combustible concentrates." The increase in the number of fires and explosions is do to the increased use of the combustible hydrocarbon propellants.

Aerosol containers are regulated by three different federal agencies: the Environmental Protection Agency (EPA) which regulates pesticide products, the Consumer Product Safety Commission (CPSC) which regulates non-pesticide products, and the Department of Transportation (DOT) which manages bulk shipments of aerosol containers. The DOT has become very effective in regulating this industry. Test methods and labeling seem to be a high priority at the DOT. But the EPA and the CPSC have not had the same attitude toward this issue. While the EPA is the major focus of this report, the CPSC virtually has actually ignored the same consumer safety concerns (i.e. proper testing methods and improved labeling). The EPA in the past has made some effort to look into this issue.while the CPSC has done little to solve the problems aerosols have caused. Ironically, the CPSC is the one federal agency that collects all of the reports of fires and explosions resulting from aerosol containers and still does nothing to improve the protection of consumers. One must remember that it is the CPSC, as well as the EPA and the industry itself that are at fault for aerosol products accidents caused by the innocence of an uniformed consumer.

The EPA began looking into the problem created by chlorofluorocarbons (CFCs) in the mid-1980's. In 1987, the EPA released PR Notice 87-8, "SUBJECT: Label Improvement Program - Required Labeling for Products Containing Flammable Propellants".

PR Notice 87-8 proposed to change the labeling of aerosol products by adding the words, "PHYSICAL OR CHEMICAL HAZARDS" to products that contain propane, butane, isobutane, and dimethyl ether as one or more of the propellants.

This statement was followed by an extended warning label that gave information about use, storage, pilot lights, electrical equipment, incineration and high temperatures. Manufacturers were given until July 31, 1988 to change labels on products released for shipment and July 31, 1989 for all products sold to the consumer.

In response to this initiative, industry dug in their heels. In an EPA memo one official states, "Industry, through the CSMA, indicated that they would initiate a lawsuit against PR 87-8 if it were not rescinded." After much debate PR Notice 87-8 was withdrawn on December 10, 1987, less than ten months after it was published. The EPA released PR Notice 87-11 rescinding the former, "based on the lack of sufficient anecdotal accident information to define the scope of the problem with respect to flammability of aerosols containing the aforementioned propellants". The aerosol industry cast enough uncertainty and the issue to terminate the proposed changes. The termination of the proposed labeling was to be followed by an extended period of data collection and research by the EPA and the industry.

Why has the aerosol industry taken such a stance against the EPA and the consumer? A major reason for the aversion to the previously proposed changes revolves around profits. Changing testing methods and labeling will cost the industry due to the redesigning of aerosol products. Changes would alter the current way the production process occurs and the industry would have to brar the cost of the new labels. Prior labeling would have to be replaced in a specified period of time to be in compliance with regulations and to avoid any penalties.

An EPA memo described a conversation an agency official had with an official from S.C. Johnson & Son, the manufacturers of Raid. According to the EPA official, after discussing testing methods, the industry official stated, "his company does not need to worry about the effects of flammability products since its insurance company settles on any damages or injuries." This statement was also verified by an internal EPA Office of Pesticide Programs memo, "Industry does not appear to be concerned as their insurance companies pay all claims."

How many fires and/or explosions are enough to cause the industry to change its approach to the use of aerosol products? So far it appears that the industry relies on the excuse of "consumer misuse" as the primary reason for accidents caused by aerosol products.. Industry should ask itself why this abuse occurs and how the misuse can be prevented. Instead the manufacturers keep trying to find possible ways to get around the system. According to EPA officials, ways of getting around the system are known, "With current approved testing methodology, a 'nonflammable' rating of the hydrocarbons and ethers can be achieved by proper engineering of the nozzle and the delivery system." Montford A. Johnsen, Technical Editor for Aerosol Age magazine, agrees with this statement. "The nature of the flame projection test invites the formulator to make modifications of product or package to pass it and thus avoid labeling the product as 'Flammable'." Yet consumer misuse is industry's answer to the problem, even though they have been notified of numerous accidents that have resulted because of an uninformed consumer.

Why would the aerosol manufacturers contimued conduct business in this manner? The propellants in question (propane, butane, isobutane and ethers) are some of the least expensive propellants on the market. Alternative propellants would cost more which could eventually drive up the final cost to the consumer.

By providing a full explaination of the contents and the explosion/fire potential, the consumer might become more sensitive to the product and its inherent dangers.For the last eight years the EPA has tried to re-educate the public about the safety risks that could save lives and property. We believe changes are needed before the accident databases overflow with victims. This process may cost aerosol manufactures a small amount at first but in the long run it will save lives.

However, it should be industry's responsibility to educate the public if they cannot use existing technology to make their products safer. The problem is that technology is available that will make the product safe. Therefore, the exisitng products would be considered defectively designed.

Aerosol Propellant for TRF Insecticide and Health Hazards

In addition to the propellant system for TRF insecticides being extremely flammable and explosive, users are also exposed to other health hazards. Exposure to the propellant, once it is released, can cause drowsiness, have an anesthetic effect, cause eye burns and skin burns (frostbite), central nervous system depression, asphyxiation, headache, nausea, unconsciousness, rapid breathing, visual disturbances, muscular weakness, tremors, narcosis, eye damage, mild intoxication, loss of coordination, airways dysfunction syndrome, difficulty in breathing and death. The target organs are the central nervous system, lungs, eyes and skin.

Since the vapor density of the aerosol system is much heavier than air, the propellant tends to stay as close to the floor as possible. When that occurs, there is a lack of oxygen or air in the area adjacent to the floor and no insect is able to survive in that environment. Therefore, contrary to the claims that the defendants made to the EPA that the propellant is considered an inert ingredient, it is very effective in actually suffocating insects on contact.

The use of the propellant system in killing insects by suffocation was not considered by any of the defendants. This is a serious omission and should have been divulged in their EPA application. Furthermore, all of the testing on rats and other animals should have incorporated the use of the propellant rather than using a non-pressurized formulation. As stated in another section of this report, this is further evidence of why the testing performed by the defendants on animals was defective. Their application to the EPA for approval of the efficacy of the active ingredients.was not complete and, therefore, defective as to the results reported and/or not reported.

Asphyxiants, the TRFs and Risk Analysis

Propane, butane and isobutene, for example, are asphyxiants. There is no question that the lower flammability limits of propane and air was exceeded causing both an oxygen-deficient and explosive atmosphere. Butane is also heavier than air as well as isobutene. Inhalation of this combination of gases may cause rapid respiration, dizziness, fatigue and nausea. The manufacturers could have looked up these symptoms on the Internet. There is no question, in the opinion of this author, that they knew of the chemical, physical and health hazards associated with the propellant they used.

In addition to the above, the undersigned consulted on a litigated case in Florida about fifteen years ago where a professional applicator using a similar product wound up with neurological and respiratory injuries. The records for this case were unavailable to the undersigned.

Prior to creating the TRF insecticides, the manufacturers knew or should have known that the standard of care, and custom and practice in the chemical industry that involves hazardous products is that the manufacturer performs a risk analysis on the foreseeable uses and misuses of their product prior to placing it in the stream of commerce. If there is nothing available from a technical standpoint that would have made that product safer or reduce the risk to the user, then they had a duty to warn and place the user on notice of the inherent hazards and dangers associated with their product and methods of protecting themselves and others from foreseeable and possible harm. Based upon the documents reviewed by the undersigned, there is no evidence that the manufacturers performed a risk analysis and properly warned users against the foreseeable dangers and hazards.

If the manufacturers had performed a risk analysis, they would have known how to eliminate the fires and explosions that have regularly occurred over the last four decades. In almost all cases it was just a matter of time, from a statistical basis that the fires and explosions would occur. This is well exemplified by the thousands of explosions that have occurred over the last several decades up to the present time.

If they had researched methods of reducing the risk of the user being exposed to fires and explosions, they would have known that there were safer alternatives available.

Consumers

Consumers were expected to know that the propellant had gone after a certain period of time. This is not within their ability to do so based upon the physical and chemical nature of the product and the environment that the product was used in. The consumer would have had to use an explosion proof exhaust system placed at the doors with all the windows open in order to remove as much of the explosive propellant as possible. Because there are areas in their homes in which the propellant may be trapped, only a professional would know how to safely remove the explosive vapors. A professional would also be able to measure the concentration of the trapped or isolated propellant in any area of the home. The manufacturers are fully aware of thar fact but not the consumer.

The CPSC

The CPSC and litigated cases document the fact that these formulations explode and injure individuals. Through the use of the Freedom of Information Act, one can obtain the incidents without identifying the individuals or parties involved. However, the reports on the Internet estimate that there are over 500 fires and explosions in the United States. Further, the pesticides in bug bombs are also potentially toxic to humans and pets. Rid-a-Bug is fully aware of those reports. The EcoSMART website reports that there are at least 500 fires and explosions resulting from the use of bug bombs. That website was also available to every manufacturer and their management.

When an accident occurs involving a fire or an explosion, often the manufacturers of that product do not make a concerted effort to determine what occurred until litigation is initiated. It is at that time that they blame the plaintiff for misusing their product.. It should also be noted that when there is a fire or explosion where there is minimal damage or injury, those incidents are very seldom, if at all, reported.

In addition, there were no documents available whereby any of the defendants reported accidents and injuries as required to the CPSC. According to their own paperwork, they relied upon self-reporting by those injured. Since the late 1980's that has been unreliable and not a proper measure of the actual accidents occurring with the use of their products.

The defendants in this case have known for decades of reports of fires, destruction of entire apartment buildings (Salt Lake City, September 2009), explosions and burned victims resulting from exposure to propellant systems used in hair sprays, room deodorizers, repellent sprays for mosquitoes insects and other similar products. The only difference is that the other products use very little of the propellant for each application.

In addition to fires and explosions, there are also serious side effects affecting one's health. Consumers have experienced coughing, shortness of breath, airway dysfunction syndrome and upper airway irritation. Insect bombs usually contain pyrethroid and/or pyrethrins as active ingredients in addition to the extremely flammable and explosive aerosol propellants. Not one of the products is natural and in some cases of injury and death, the users followed instructions and vacated their premises for hours. In spite of this, there is one known death from insect bombs involving an infant found dead the day after the insect fogger was used in the infant's home.

The fact that the manufacturer did not list the whole formulation including the propellant system as part of the formulation in their application to the EPA and the public is, in my opinion, misleading and fraudulent. That EPA was never informed of the highly flammable and explosive nature of the propellant system in any of the applications. However, manufacturers, in their applications using the non-flammable propellant, made sure that the agency knew that it was "non-flammable." Their statements and applications speak for themselves.

It is a known fact that the courts do not recognize the self-reporting of injuries because most companies do not make an effort to know what occurred with each and every TRF that was sold. In addition, there is no evidence that manufacturers have reported any incident involving their products to the CPSC. In many instances, the lawyer representing the injured party is the one who makes inquiry.

Warnings and Instructions

The warnings and instructions placed on the subject products are defective. It is difficult to believe that the defendants claimed federal exemption when they misled the EPA with incorrect and fraudulent data regarding the extreme flammability and explosive characteristics of the Total Release Indoor Aerosol Insecticide.

The product does not include hang tags that incorporate the phrase **"DANGER DO NOT RE-ENTER UNTIL THE ROOM/AREA IS COMPLETELY VENTILATED WITH NEW AIR AND ALL FLAMMABLE/EXPLOSIVE GASES ARE ELIMINATED."** This tag should be placed prominently on the entrance to a room or house restricting entry until the area bombed is completely aired out. The hang tag statements would be in both English and Spanish.

In addition to the above, the warning on the subject product should also place the user and others on notice that the product is too dangerous to be used by anyone other than a professionally trained person. The warnings should include the following:

!DANGER!

THIS TRF FOGGER IS EXTREMELY FLAMMABLE AND EXPLOSIVE

THE PRODUCT MUST BE APPLIED BY PROFESSIONALLY TRAINED AND CERTIFIED TECHNICIANS

DISABLE ALL ELECTRICAL PRODUCTS, HOT-WATER HEATERS, FURNACES, REFRIGERATORS, STOVES, PILOT LIGHTS, SPACE HEATERS, OTHERWISE AN IGNITION SOURCE WILL IGNITE THE EXTREMELY FLAMMABLE AND EXPLOSIVE INVISIBLE VAPORS

IF THE PRODUCT IS APPLIED BY ANYONE BUT A CERTIFIED AND TRAINED TECHNICIAN, THERE IS A HIGH PROBABILITY THAT A FIRE AND/OR EXPLOSION WILL OCCUR RESULTING IN SEVERE BURNS AND POSSIBLE DEATH

AFTER 5 HOURS OPEN ALL WINDOWS AND DOORS AND APPLY EXPLOSION PROOF EXHAUST FANS FOR AT LEAST ONE HOUS TO MAKE SURE THAT THE EXPLOSIVE VAPORS ARE ELIMINATED

In addition to the above, a pictogram of a flame and explosion should also be placed on the label. The requirements for the pictogram can be found at 40 CFR 156.78(d).

Another condition that is not mentioned but is an important factor is the weather. In extremely cold climates, is difficult for the product to be used because one would have to open up the windows and doors of the residence, apartment or house, in order to completely eliminate all of the extremely flammable/explosive propellants. The problem is that the gaseous propellant is invisible and only a trained certified technician would be able to measure and determine whether all of the propellant has dissipated and is eliminated.

The Home Insect Control--The Manufacturer's Hype

Dr. Michael F. Potter of the Department of Entomology, College of Agriculture, University of Kentucky, reported in the paper titled, Limitation of Home Insect Foggers ("BUG BOMBS") that, "while foggers are convenient and require little effort or knowledge to dispense, they seldom are effective against indoor pest problems." He goes on to report that "when applied in this manner, very little insecticide actually penetrates into cracks, voids, and other hidden locations where cockroaches, ants, silver fish, and most other household pests congregate and spend most of their time. The ingredients within "Bug Bombs also tend to be repellent, causing insects to scatter and move deeper into wall voids and other hard-to-reach areas."

It is a known fact among entomologists that when insecticides are needed to eliminate pest infestation results are much better if a spray, bait, dust, etc., is placed directly into the areas where the pests are likely to be hiding. They have found that targeted insecticide applications are much more effective against all varieties of household pests, including fleas.

In addition, when using TRF insecticides, the resulting pesticide residues wind up on exposed surfaces and constitute a health hazard. Occupants with asthma and other respiratory ailments react severely when, for example, pyrethrins and other irritating volatile compounds are used indoors.

Safer Alternatives

The technology for safer alternatives has been known for decades. In fact, room deodorizers and hair spray accidents involving fires and explosions using similar propellant systems as the TRF insecticides have been reduced significantly by changing the propellants to non-flammable gases or with the use of manual sprays.

DuPont's Dymel, as previously is a non-flammable propellant system that the defendants knew about and actually applied for an EPA registration approval of the same formulation as the subject product in this case using the DuPont propellant in the year 2000. In spite of this, they continue to use the highly flammable and explosive propellants which ended up severely burning consumers and destroying property.

It should be noted that several or more of the major manufacturers made application to the EPA using DuPont's Dymel 134A. They identified the actual propellant as part of the total formulation. Dymel, although toxic like the combination of extremely flammable and explosive ingredients used in the subject product, is non-flammable and inert. Major manufacturers all knew that this product was available and could be used safely by the general public.

Spectracide, distributed by Spectrum Group, a Division of United Industries Corp., manufactures a non-flammable insecticide that kills ants, roaches, spiders, fleas and ticks. In the EPA registration # 9688-176-8845, the active ingredient is lambda-Cyhalophrin. This product is disbursed by manual spraying and it more than documents the fact that manufacturers using the propane/butane/isobutane propellant had safer products available that would have eliminated the risk of fires and explosions.

The ORTHO Group manufactures and distributes an insecticide called Home Defense Max. The product is non-flammable and will not explode when exposed to an ignition source. The product kills ants, roaches, spiders, fleas, ticks, earwigs, silver fish, crickets, carpet beetles, firebrats and moths. The EPA registration number is 239-2663. The active ingredient is bifenthrin.

Hot Shot Pest Control Concentrate is also a non-flammable, one-ounce product and can be diluted with water to create up to two gallons. The active ingredient is opernethrin and the EPA registration number is 73049-154-8845.

It might suggested that selling TRF insecticides in smaller sizes and as individual "bombs" would reduce the risk. Is a smaller "bomb" safer than a larger "bomb?" The answer is definitely "No." The consumer is still disbursing the explosive and extremely flammable propellant. The smaller size will not significantly reduce the risk.

Conclusions

FIFRA §3(d)(1)(C) provides for an insecticide to be classified for restricted use. The EPA is concerned about unreasonably adverse effects on the environment.

EPA regulations at 40 CFR 152.170 present a more detailed criteria for making a FIFRA §3(d)(1)(C) determination for an end-use product. When deciding whether to classify a product for restricted use, the Agency considers the toxicity of the product along with other pertinent evidence of hazard and, if any serious hazards could reasonably be mitigated by restricting use. They examine whether labeling could adequately address those hazards. Many manufacturers do not divulge all of the risks and hazards associated with the product which is, in this case, the extreme flammability and explosive nature of the propellants used. The fact that many manufacturers made false, misleading and fraudulent statements in their application to the EPA including all of their testing data, should eliminate any claim they would make as to being federally exempted in their product labeling. It is obvious in all cases involving their EPA applications that they intentionally failed to place the EPA on notice by identifying the "inert ingredients" as being non-flammable as described above.

In addition, safety should never be an option with reference to a product being federally exempted. Manufacturers intentionally compromised the safety and welfare of consumers when they could have easily used the non-flammable propellant systems available for their product. Manufacturers should never place TRF insecticides using highly flammable/explosive propellants in the stream of commerce where laypeople can purchase it. They are not trained in the safe use and application of a highly flammable and explosive product.

The EPA was never given the opportunity to weigh the seriousness or magnitude of the relevant risks in order to consider whether restricting the use would be an appropriate means of reducing those risks. The benefits do not outweigh the risks when there are safer alternatives available to the general public.

Author Information:

C. J. Abraham, PhD, DSc, JD, P.E., FRSC, FTI, DEE (IH), BCEE has spent the last 40 years consulting internationally on a variety of personal injury cases. He has also invented and commercialized a variety of products. Over the last 40 years he has evaluated many household and industrial products with reference to their foreseeable uses and foreseeable misuses. He has also tested a large number of products for their flammable and explosive characteristics and created a number of warnings and instructions as well. He is certified and has diplomat status in a number of areas. His background and experience can be found at www.scientificadvisory.com.

(36d) Tips From 25 Years of Environmental and Product Liability Litigation

Gayle S. Koch, Axlor Consulting LLC, Cambridge, MA

Chemical engineers, by choice or chance, often become involved in litigation proceedings. Experience with various environmental and product liability litigation leads to lessons on how to be aware of litigation issues and better prepare for the experience. This session will cover concepts such as economic benefit, allocation of costs, quantification of costs under uncertainty, cost causation, disclosure, and risk assessment issues. In addition, general tips such as standards, documentation and communications will be discussed.

"Innovative Project Management for Large Scale Process Projects"

Stephen L. Cabano
President, Pathfinder, LLC

Introduction

When we look at today's Oil & Gas Industry, we see projects that dwarf anything we have executed in the past. LNG projects in Australia and elsewhere are running in USD$10 - $50 billion range. Petrochemical companies in the Middle East are evaluating projects which represent USD$20 billion in capital investment. Offshore Oil & Gas projects with initial estimates of USD$5 to $10 billion are no longer "outlier" or uncommon data points in some companies' project portfolio planning considerations. In addition to the sheer magnitude of capital cost, these projects also face a myriad of technical and execution challenges, as companies stretch their technological, logistical and human resource expertise capabilities to new limits.

Mega-scale projects have also become "global operations" and enterprises based on the need to support project sites in various, often remote, regions around the world. The scope and magnitude of these projects often necessitates the use of multiple engineering firms, procurement sourcing, fabrication facilities, etc., as these critical resources are pressed to the limit of their capacities in any one location/office. This unprecedented expansion also comes at a time when the very resources we need – particularly in terms of human resources and expertise, are in short supply, with the trends indicating an ever dwindling resource pool of experienced practitioners in virtually every technical and project/construction management discipline.

Facing this mountainous combination of opportunities and associated challenges, Sponsors of these initiatives need to seek out, develop and implement innovative execution approaches to address the realities of promoting and successfully bringing these global, mega-scale project plans and programs to fruition in the most economically feasible and profitable manner, and on a schedule that supports supply and demand projections. Typical considerations include effective and realistic planning at more detailed levels in early stages of project/program development. These planning requirements also include examination and selection of the most effective contracting and procurement considerations, emerging/"mature" technology combinations/selection, techniques for resourcing the team, efficient and effective construction techniques, coordination and transportation logistics, construction labor resourcing and capability within the host region. Host country requirements may stipulate local labor content, which, in some cases, may require ~80% of work force sourcing from within that country or region. These are but a few of the requirements a mega-scale project sponsor must address.

Many of these mega-scale projects are being developed by or include nationally-owned or quasi-government ownership. Thus, national interests must also be considered and integrated along with the process/production issues involved. In addition to local professional and craft labor content, requirements such as extensive infrastructure development requirements (e.g., roads, schools, and hospitals), other social/quality of life aspects, unique environmental sensitivities, etc. may be imposed upon the project and will need to be addressed as part of the overall program.

Planning, scheduling and estimating projects of this size, scope and complexity can also prove to be particularly challenging as traditional metrics need to be adjusted for scale, interface, logistical and regional considerations and dynamics. The sheer magnitude of these projects means that multiple organizations will likely be engaged in the development of the engineering and construction work packages associated with the chemical processing and/or production units and plants included in the mega-project scope. This can mean that literally thousands of such work packages need to be planned, developed, scheduled, coordinated and structured to align with construction resource capabilities and capacities, as well as facility Commissioning and Start-Up requirements. While these detailed packages are an absolute necessity, the roll-up/roll-down integrity must also be absolutely respected and maintained in order to provide the right level of detail to the specific end-user involved. While Construction Contractors and Sub-Contractors require detailed packaging, Sponsors and Owners require summary level information digestible at executive levels to assure informed, strategic decision-making with respect to financial commitments and other project direction needs.

Recognizing that Owner expertise may be more aligned with entrepreneurial skills and expertise rather than in project implementation, many of these enterprises have turned to utilization of Program Management Contractors (PMC) who have more extensive expertise in managing project-related coordination and implementation challenges. This poses a challenge for even some of the largest Engineering/Procurement and Construction (EPC) firms as their own workload consumes much of their resources, particularly those which are most talented/experienced and possibly already assigned to more than one such mega-scale program. In many cases, this necessitates formulation of consortia or partnerships with possible "competitors" to acquire the span and levels of resources needed to provide Sponsors/Owners of these mega-projects with a single-source of program management and/or technical expertise required to fill such roles effectively. Even this approach, although successfully applied in the past, has its own pitfalls and challenges to overcome within the context of the mega-project environment, not the least of which may include protection of intellectual property, trade secrets, company confidential information and business techniques.

The challenges noted above may appear to be monumental and extremely
difficult to manage and control. However, since the opportunities involved offer the promise of profitable (often socially attractive) outcomes, Sponsors and Owners find it increasing difficult to walk away from such enterprises, and, as such, continue to seek out the best alternatives and techniques available to address and mitigate the risks inherent with such large-scale endeavors. This paper will address many of these challenges and provide insights into proven/tested approaches, techniques and execution alternatives applicable to the mega-project environment.

The Project Management Approach

The following diagram (Figure 1) shows a generic, five-Phase Project Delivery Process (PDP), similar to that used by many Owner organizations involved in process industries:

OWNER PROJECT DELIVERY PROCESS

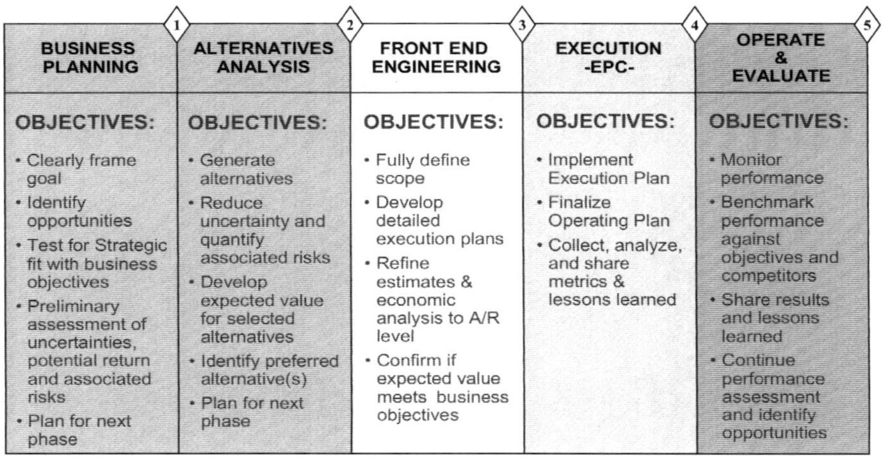

Fig. 1

This project approach to project Planning and Delivery will form the basis for all discussions moving forward.

1) Integrated Project Delivery Processes, Guidelines and Procedures

As previously noted, many mega-scale projects involve multiple Sponsor and Owner entities, and may also involve a PMC organization as well as a number of EPC firms working together to develop and implement the mega-project. While these companies typically have well-developed PDPs in their own right, if the mega-project involves two or more sponsoring organizations, a PMC and multiple EPC firms, it becomes necessary to either adopt one of the participant company's processes and underlying guidelines, or to carefully examine the system to select the "best of the best". This technique may also be applied to General Engineering Standards and Specifications.

If such a selective approach is used, it is incumbent upon the parties to also ensure that the possible conflicts or inconsistencies between the selected processes, standards and procedures (both technical and non-technical) are adequately "scrubbed" to avoid confusion and unnecessary recycle to correct errors and/or omissions during application.

2) Front End Loading Process

The essence of Front End Loading (FEL) is discipline – the discipline to examine the project in detail, starting with the very earliest of its development stages to ascertain its feasibility and risks. If the opportunities (confirmed through thorough and realistic techno-economic analyses) outweigh the risks, the project is usually authorized to proceed to next step in the PDP.

In any project, but more importantly in mega-projects, due to the level of risk and investment involved, deliverables such as technological and engineering studies, site selection alternatives, marketplace conditions and forecasts, environmental and socio-economic requirements should be progressively examined and used to guide development of estimates, schedules, and execution strategies and plans.

Although PDP and FEL processes have been criticized by some as being too restrictive, confining or stifling with respect to "outside-the-box" thinking, these processes and practices play an important role in decision-making, particularly in light of the enormity of capital investment, time and resources involved in a mega-scale project. As such, careful application of FEL principles and practices becomes an even more important aspect of Sponsor/Owner due diligence in a mega-project situation. This discipline takes on a vital role in situations where multiple decision-making parties are involved and transparency is of paramount importance to achieving alignment and buy-in across the organization.

Ensuring adherence to FEL processes and practices, including in-depth Stage Gate and associated deliverables reviews, also assures that cost estimate and schedule premises are based more on objective information, rather than on subjective data, enhancing the opportunity for achieving required accuracy levels. This further builds confidence among decision-makers, lenders and other stakeholders to the soundness of the mega-project.

3) Project Charter and Project Execution Plan

The importance of clearly defined Business and Project Objectives in a mega-project environment cannot be over-emphasized. These Best Practices (and the documents used to convey these over-arching principles and practices) should be documented in the Project Charter to facilitate understanding of the desired/required outcomes, and the means by which to achieve those goals. The Project Charter often takes the form of a separate document or, just as often, is incorporated as the "preamble" to the Project Execution Plan (PEP).

First and foremost, Business and Project Objectives should be stated in clearly understandable and measurable terms. Business Objectives should also be clearly "traceable" through to, and identifiable within, the Project Objectives. Use of SMART principles is highly recommended. As such, Business and Project Objectives should be stated in terms which are:

- **S**pecific
- **M**easurable
- **A**ggressive but Achievable
- **R**esource Dimensioned
- **T**ime Dimensioned

The Master Project Execution Plan (MPEP) should clearly define the 'what, when, why, where, how, how much and by whom' dimensions, providing the "roadmap" for the methods and means for achieving the Business and Project Goals and Objectives. In a mega-project environment, the MPEP should cover the overall program. Each individual sub-division of the mega-project (Upstream and Downstream, Process Units, Off-sites & Utilities, etc.) should have a more detailed, unit/package-specific PEP developed which uses the MPEP as its backbone but addresses any authorized deviations from the norms included in the MPEP and/or describes any particular nuances associated with that particular piece of the mega-project. Whether at the

MPEP or individual unit/package PEP level, MPEPs and PEPs should also take care to integrate each of the topical coverage areas to ensure consistency, coordination and compatibility across the full spectrum of strategies and plans included in the document. Likewise, it is equally important to describe interfaces between each package and its relationship to the overall program.

An example of the content and coverage which may be considered in a mega-project PEP is include as Figure 2 below:

PEP Table of Contents

Section 1 - PROJECT CHARTER

- Strategic and Business Objectives
- Project Objectives
- Scope of Work
- Support of Local Communities/ Workforce/ Enterprises

Section 2 - PROJECT PLANNING BASIS

- Risk Management Plan
- Organization Plan
- Master Plan/ Schedule
- Contracting Plan
- Funding and Cost Estimate Plan

Section 3 - PLAN FOR IMPLEMENTATION

- Design Management and Value Improving Practices (VIP) Plan
- Constructability Plan
- Procurement and Materials Management Plan
- Construction Management and Labor Plan
- Mechanical Completion/Commissioning Through Start Up Plan
- Regulatory and Environmental Management Plan
- Project Controls Plan
- Quality Management Plan
- Safety Management Plan
- Informational Management and Communication Plan
- Document Management Plan
- Operations Management Plan
- Plan to Capture Lessons Learned
- Special Factors Plan (Modularization, Project Suspension/ Restart Plan etc.)

Fig. 2

It should be noted that the example included above is not intended to be all-encompassing in terms of the content and coverage for all mega-projects. It should instead be considered as a minimum requirements guideline, subject to modifications based on the mega-project specifics, its characteristics, partner/stakeholder requirements, etc. While the actual content and coverage may vary, the common denominator in the development of the Project Charter and Master & Individual PEPs is communication of a uniform message to ensure full understanding and alignment of the goals and the path to achieve them across the enterprise. Given the dynamics associated with mega-project environments and the time needed for decisions to be disseminated throughout the enterprise, it is imperative that the Charter and MPEP/PEPs be maintained and available in real time, rather than simply created as a matter of course.

4) Scope Development

Although mega-projects are conceptualized as overall, large scale initiatives (which in most cases they are), in reality they are a cohesive and integrated collection of various process and support units, which, in and of themselves would be considered significant projects (and capital investments) in a stand-alone mode. It is therefore logical and reasonable to consider these packages as discrete scopes of work within a greater whole (the mega-project). As noted earlier in this article, some of these stand-alone packages or pieces would present a significant challenge to accomplish on their own in today's economic and resource-constrained environment.

Many companies have therefore addressed this issue by breaking down their mega-projects into "Manageable Pieces" that correspond to the capacities and capabilities of the Contractor firms (whether EPC, Engineering or Construction only) which they have identified as the best suited to accomplish that piece of the of the mega-project Scope.

In addition to this carefully orchestrated and thoroughly planned out Work Breakdown Structure of manageable pieces, it is imperative to assure that all appropriate deliverables have been adequately developed to levels of detail and definition in conformance to PDP and Stage Gate requirements. Sponsors/Owners and their contractors have a number of techniques available to assure such compliance. Among those most commonly used as independent assessment tools are the Front End Loading Index (FELI) developed by Independent Project Analysis (IPA) and the Project Definition Rating Index (PDRI) developed by the Construction Industry Institute (CII).

In addition, Scope optimization, to assure alignment with Business & Project Objectives, on a "no more, no less" basis, requires the application of appropriate Value Improving Practices (VIPs), such as Class of Facility Quality, Process Simplification, Traditional Value Engineering, Constructability Reviews, etc.

5) Organization and Staffing

Operating on the axiom that people, not companies, build projects, it is vital that the team developed to manage the mega-scale project be of the highest caliber. This starts with the Leadership Team. The most successful mega-projects have been carried out under the direction of effective leaders with extensive experience and pragmatic/realistic perspectives regarding mega-project risks and challenges. These leaders have learned the value of coaching, mentoring and guiding their teams by example and also delegating responsibility, accountability and commensurate authority to qualified members of their team. Clear, well-defined, hierarchal reporting lines, open and candid communications, and unambiguous roles and responsibilities are key tools used by the most effective project teams.

As previously mentioned, effective leaders and teams are confident in their abilities, but also cognizant of their limitations. Therefore, they seek out and secure qualified surrogates with the necessary skill sets (whether individually or through contracting with capable/experienced firms with the required skills) to complement/supplement in-house competencies and resource "gaps".

6) Risk Management

While many of the "tools" and techniques already cited in this article may be categorized as Risk Management measures, there is no substitute for a robust, well-managed Risk Management Program that incorporates a carefully maintained, updated and vigorously pursued follow-up protocol. This includes but is not limited to a Risk Register and Lessons Learned analysis regime. These tools, coupled with realistic Risk Analysis of Cost Estimates and Project Schedules, help guide the team in identifying, assessing & mitigating risks, and capitalizing on opportunities. This results in a structured set of contingencies and back-up plans, ready for implementation to address issues as they emerge (before they become "problems") – proactively, rather than reactively.

7) Contracting & Procurement Strategy

Contracting & Procurement Strategies and Plans must also be carefully considered and structured to assure that the most qualified partners for the mega-project sponsors are selected and completely aligned/supportive of the Business and Project Goals & Objectives. The screening, pre-qualification and qualification processes need to be designed to address key issues such as experience, capability, capacity and performance on previous projects. From a procurement perspective, some Sponsors/Owners have developed alliances or partnerships with suppliers to assure consistency and standardization across the entire mega-project. This provides suppliers with the confidence that "stockpiling" materials and mobilization for fabrication of equipment and materials will provide a stable outlet, with a known base load of committed work. Sponsors and Owners also tend to benefit from volume pricing structures, reduced spare parts inventories and, in some cases, a single, dependable source of supply which reduces the need for extensive (and possibly expensive) logistical coordination in terms of expediting, inspection, transportation, customs clearance, warehousing and other management challenges.

8) Project Controls

As most mega-projects typically include multiple contractor organizations, Sponsors/Owners also encounter a multitude of Project Controls systems utilized by these Contractors. These systems, in turn, are expected to feed the Owner's or PMC's Project Controls system with accurate information in a manner which can be assimilated and consolidated seamlessly within the Sponsor/Owner/PMC's system to permit effective management of the overall enterprise. This cannot and will not occur unless the sponsoring organization defines and mandates, in very specific terms, what is required by the managing organization to effectively execute the Project Controls function.

This includes defining in each Request for Proposal (RFP), the specific deliverables, the reporting sequence cycles, forecasting requirements, etc. During the bidding process, the managing organization needs to evaluate the Contractors' abilities to deliver these requirements and include them in the final Contract. In pre-qualifying, qualifying and selecting their Contractors, most Sponsors/Owners are careful to examine a given Contractor's Engineering, Procurement and Construction experience, expertise, capabilities and capacities in detail – including the caliber and quality of their systems, processes, procedures, tools, and most importantly, their people. When it comes to the Project Controls function, however, this diligence is not always applied to the same degree. In fact, most RFP's and contracts relegate the details of the Project Controls function to Post-Award discussions. Considering the number of functions involved in a typical Project Controls system (as shown in Figure 3 below), leaving such an important aspect of mega-project management to chance has grown to be a significant risk in and of itself.

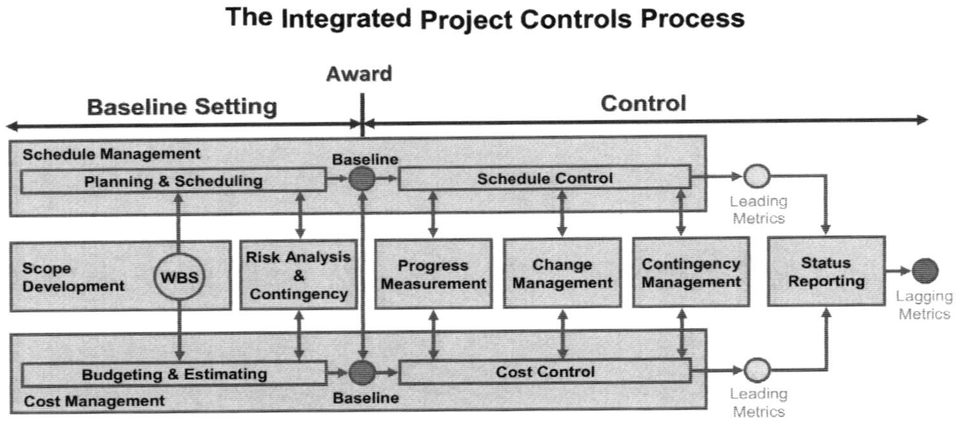

Fig. 3

It is therefore strongly recommended that Project Controls capabilities and the quality of the personnel assignment to that function be thoroughly vetted and demonstrably tested prior to Contract Award with equal thoroughness and diligence as that afforded the Contractor technical solution itself. Similar assurance measures should also be applied to the Sponsor's system – particularly if the system is based on a collection of various pieces drawn from the Sponsoring organization and its partners and/or unproven in practice.

9) Integrated Plan/Schedule and Work Face Planning

Construction of the mega-project, particularly if sited in a remote location, is most often one of the highest risk factors to cost and schedule over-runs. To mitigate these risks, mega-project Sponsors/Owners have employed combinations of modularization, detailed Constructability reviews, and Work Face Planning techniques to thoroughly analyze and plan the Construction effort well in advance. These processes include integration of construction expertise with early design efforts to assure that appropriate levels of detail, engineering deliverables, procurement cycles for equipment and materials deliveries, and on-site & off-site marshalling areas are properly coordinated.

Scheduling of this also requires careful orchestration and coordination using "backward pass" integration beginning with the commissioning and Start-up sequences and milestones to drive Construction priorities, which in turn drive Procurement and Engineering priorities, sequences and start/completion requirements.

10) Orientation & Training

Finally, since the mega-project organization, from Sponsor/Owner to Contractor/Sub-Contractor entities are unlikely to have ever worked together in an enterprise as a "complete" organizational assembly before, orientation and training in the inner workings (and each "player's" roles and responsibilities within the enterprise) is an absolute necessity. While the terms 'team building' and 'alignment workshops' may have become over-used catch phrases which have lost their true meaning, it is imperative that these techniques be applied in the vein of their original intent – assuring understanding, coordination and cohesion.

As with any new team, the various organizations within the mega-project must learn to work together as efficiently as possible, accelerating through the "forming, storming and norming" phases and settling into the performing stage as quickly as possible. This does not just happen by itself. It needs to be deliberately fostered, guided, nurtured and maintained through leadership, guidance, commitment and consistent communication.

The best method for demonstrating commitment to the teams' success and assuring open channels of communications is for the Sponsor/Owner to provide orientation sessions that enable/facilitate understanding of common goals and objectives, and the means by which those goals will be achieved.

Conclusion

There is an axiom which states, "Those that do not study history are destined to repeat it." Often, we hear that a mega-scale project is "first of kind", "unique", "paradigm- shifting", or a "step-out" initiative. Very often what's implied is that traditional practices, procedures and approaches do not apply. Time and again, this has proven to be a risky approach/perspective. While it is true that many mega-scale projects are 'pioneering' endeavors – proven and accepted Best Practices and Lessons Learned should not be discounted or ignored. Rather, these Best Practices and Lessons Learned should be considered as the core principles and practices, from a starting point perspective. These core values should then be brainstormed and analyzed as to how the benefits that defined them as "Best Practices" and "Lessons Learned" in the first place, may be extracted in a more innovative and/or customized manner to align with the unique characteristics of the mega-project in question.

(68b) Sustainable Development Pays: Financial Success Is Rooted in the Ability to Identify and Respond to Key Social and Environmental Issues

Dana Cooper, Fairmount Minerals, Benton Harbor, MI

Sustainable innovation can drive the success of a business. Sustainable development affords a social license to operate and helps to drive innovation and growth. As a business, profitability creates prosperity for our shareholders and employees while contributing to economic development in our communities.

Enhancing the sustainable attributes of our products by utilizing sustainable materials, such as "green" resins or recycled content, and ensure proper marketing of the products in a way that is in compliance with relevant regulations and codes. The stronger the bottom line, the more positive direct economic impact. From innovative waste elimination to energy efficiency to wellness initiatives and more, we find that sustainable development pays.

(68c) Sustainable Business Operation and Environmental Preservation - the Evolving 21st Century Thinking on the Achievability of Both

Jacqueline Sibblies, Environmental Compliance Consulting, Somerset, NJ

The governmental implementation of environmental protection measures of the 1970's generated passionate debate among stakeholders on the results that these effected procedures could or would make. A large section of the general public and several public advocacy groups felt that requirements were strict enough on the industry to bring about the kind and degree of operational changes needed to appreciably mitigate the state of environmental deterioration. Members of the industrial sector viewed the environmental protection measures as very restrictive, undermining business profitability and possibly threatening to business survival. The regulatory agencies considered the requirements reasonable and adequate in addressing the then prevailing environmental concerns.

The arguments and diverse perspectives on environmental protection requirements have persisted over the decades, intensifying at pivotal junctures. As scientific research and human health evaluation data continue to reveal increasing and/or new negative environmental effects from industrial operations, governmental agencies have sought to reduce these factors through further regulations, either by amendments of existing regulations or promulgation of new ones. Such moves have left industry with a heightened sense of being subjected to overly burdensome requirements that could not be complied with economically. Some thought that there were requirements that were simply not necessary and that industry was in the best position to devise procedures for controlling practices that negatively impact the environment, should be allowed to develop appropriate countermeasures.

Indeed by the 1980's, the chemical industry embarked on a set of self derived standards of operation known as Responsible Care, developed in Canada. Participation is voluntary and requires that members follow recommended practices of safe handling and disposal of chemicals, monitor these practices and keep records of specified metrics. Later, the European non-governmental organization (NGO) International Organization for Standardization (ISO) developed standards for an environmental management system which included requirements for regulatory compliance.

Despite these efforts, the 1990's brought even more stringent environmental regulations due to demands for a cleaner environment and growing concerns regarding climate change by the public, public advocacy groups and environmental groups. Internationally, consumers began to indicate their preferences for goods and services that incorporate environmental preservation. Members of the industrial sector started to look for way in which they could fulfill consumer expectations and still maintain acceptable business returns.

During the first decade of the 21stcentury, various sectors of industry led by prominent leader in those sectors developed operational/business models that have succeeded in sustaining high environmental standards and economic performance. This paper will discuss the development of the new thinking that generated this industrial paradigm shift and offer a projection into an era of fuller stake holder collaboration towards sustainability.

Identifying Discontinuities for Research and Technology Development

In the area of research and technology planning and technology development, major shifts in product and process development can occur that often catch current users or owners of an existing process or product technology by surprise. Current examples would include Kodak's late and ineffective response to electronic based photography and Borders' slow response to electronic books and on line marketing.

On the positive side, we see new product development activities integrating functionality of products to be very pro-active in such areas as oral care products, office equipment, mobile communication devices, and integrated chemical complexes that minimize energy consumption and waste disposal.

When we study the history of breakthrough inventions via the patent literature, we find a number of recurring themes that cut across all areas of science and technology that allow an accurate, but generalized, forecasting tool set that can be used to be pro-active in the identification of discontinuities. These themes, frequently called Lines of Evolution, are at the core of the TRIZ (Russian algorithm for "Theory of Inventive Problem Solving") problem solving and strategic planning algorithm and tool set.

A few of these are:

1. Systems and products become more ideal from the standpoint of customer application and resource utilization. While all chemical and process operations are constantly trying to reduce costs, this viewpoint is more aggressive. It asks us to think about not using a material at all and to actively look at the system, sub-system, and super-system for resources that could be used, or systems that could be integrated to minimize materials usage. A recent example of this in the automobile industry is the Mazda plant in Illinois that has totally eliminated its landfill. Other examples include some chemical operations that are coupled with other plants and operations that can utilize the waste heat generated by exothermic chemical reactions.
2. Increased dynamism within a product or process operation. This would include what has been done for a long time such as automatic price escalators in pricing related to a specific raw material price. It would also include the operation of various chemical plants as a function of raw material pricing and availability. A sub-set of this is the changing of various unit operations within a chemical complex to produce different varieties of materials as a function of seasonal demand (gasoline vs. diesel fuel for example). The dynamic response of gasoline blending to match seasonal weather conditions is another.
3. Oscillation between adding useful complexity vs. simplifying. In the cell phone/mobile device area we have seen what started out a device which made only phone calls slowly add more and more functionality to the point where many of these devices are replacing PC's and separate Email services. But we are beginning to also see the re-introduction of simple phones, which are only used for calling, as well as unbundled pay plans. How does this relate to the chemical industry? We have seen Dow Corning, for example, while continuing to develop specialty materials and customer service, also set

up a separate subsidiary (Xiameter™) which only sells large quantities of commodity silicone fluids with no customer service provided.

4. Upward system integration. When we analyze how systems and processes have evolved, we see a clear trend toward upward system integration. What does this mean? Let's consider

5. Field progression. If we study the progression of technology in the patent literature, as well as simply look at products around us, we see a natural progression from the use of simple mechanical fields, to thermal and acoustical fields, then to chemical fields, on to electronic fields, and finally optical and electromagnetic fields. Cooking systems and communication systems illustrate this progression graphically. In our industry, what have we seen in the area of polymerization technology? Quite some time ago, we literally heated up monomers and hoped for the best in molecular weight. Then we found catalysts and stereo specific catalysts that could control molecular weight, polymer cross-linking, and other properties very accurately. We now see exploratory research in the area of microwave catalysis. This line of evolution predicts that this will be the next step in this technology. It does not say when, just that this will be the next step. Before we dismiss this possibility, remember that microwave ovens used to cost $500 and only boiled water!

In reviewing these lines

25 Years of Process Design + Manufacturing + R&D and Still Don't Know What I Want to Be When I Grow up

Ahmed J. Allawi, Innovation R&D, Shell Projects & Technology, Houston, TX

Tuesday, April 3, 2012: 8:35 AM
340A (Hilton of the Americas)

Abstract

The author spent most of his career in the Engineering and Construction industry (E&C) where he worked in process & environmental engineering, technology support, and project & department management. After 20 years in this industry, he took a leap of faith and joined one of his client companies to manage the manufacturing support functions at a major petrochemical complex. Two years later, he moved to his present role where he leads a regional Research & Development group within the same company. The author shares his first-hand experience of working in these numerous roles in three different industries. In addition, he discusses the value of advanced degrees in engineering and/or business in each of the industries in which he worked. The author's main goal for the paper is to provide practical career guidance to young engineers as they consider their options for professional development and advancement.

Introduction

"Why would a young engineer ever want to work for an Engineering & Construction company? No one ever listens to you before you hit 50!" That was the unsolicited advice of one of my former colleagues, which he offered me in the late 1980's. At that time, I was contemplating the option of leaving a job with a small start-up company to join one of the world's most respected Engineering & Construction (E&C) firms. Although I am glad I didn't heed his advice, I promised myself I would keep it fresh in my mind so that I can offer it someday as a (possibly) dissenting point of view.

Nearly a quarter of a century later, I find myself puzzled the scarcity of practical career guidance available for newly minted chemical engineers. Much as it was in my time, advice from older colleagues still plays a large role in forming both biased and unbiased opinions on employment options. While I don't claim to be unbiased, I think I can present my experience in a way that would help someone make more informed career choices. My hope is to provide the reader with the kind of insight that will generate more questions than answers. It is only by asking these questions and answering them for yourself can you make a truly informed choice.

Early Career Choices

As far back as I can remember, I've always wanted to be a chemical engineer. Growing up in the 1970s, I witnessed firsthand what the Arab oil embargo did not just to the economies of the Western world but also to the economy of Iraq, the country of my birth. The buzzwords of the day were "Synthetic Fuels", "Coal Gasification", and "Biomass". Big rewards awaited the companies and individuals who were the first to find the proverbial Holy Grail and produce low-cost fuel from inexpensive raw materials. This social economic climate had a profound influence on our profession. Chemical engineering enrollment swelled in the late 70s and early 80s, and by 1982 it reached a peak. At my alma mater, for example, the class of 1982 included no fewer than 220 chemical engineering graduates - almost a third of all graduating engineers. What we didn't know at that time was that we were in the midst of an artificial bubble. As soon as President Reagan stopped funding for the development of synthetic fuels, the chemical engineering world started a rapid decline. By the time I finished my undergraduate training two years later, my graduating class had shrunk by one half from the peak of 1982. Two years after that, the oil bust hit, and the size of the chemical engineering graduating class was down to a quarter of its peak. The disconcerting part was not in the drop in the supply of chemical engineers but rather the underlying drop in demand: In 1980, every chemical engineering graduate had multiple employment offers, but by 1984, only one third of my classmates could boast about an employment offer before graduation.

I am sharing this background because it's important to understand the context of the career advice and/or insight you might be receiving from your older colleagues. The chemical engineering graduates of the early and mid-80s are much like the survivors of the Great Depression: Our parents, grandparents, or even great-grandparents had to adapt their lifestyles in response to what they experienced in the Great Depression. In an eerily similar way, chemical engineering graduates of my generation had to shape their careers in response to the job market of the 1980s. Many got out of engineering altogether. A sizeable contingent elected to go directly into graduate school without even bothering to look for a job. Those who were fortunate enough to land a job with a good company at that time are probably still employed by the very same company today.

My circumstances then were little different than most of my colleagues. For one thing, I was on a student visa which disallowed employment after completing my undergraduate training. For another thing, I had a strong interest in pursuing postgraduate training - with the hope of eventually embarking on a career in process design at a major E&C company. Much as it is today, the optimal degree depended, to a large extent, on what the individual wanted to eventually do for a living: A bachelor's degree for operations support, a master's degree for process design, and a PhD for Research & Development. While the prospect of pursuing a PhD was very appealing to me, I could not convince myself that a PhD would be viewed as an asset in the E&C industry. I consider this to be one of the toughest career decisions I have ever had to make. On the one hand, I immensely enjoyed not just the research work I was doing but also the thrill of working with a world-renowned expert in a field that ignited my passion. On the other hand, I couldn't see myself doing the same line of work in a laboratory environment for a good part of the next four decades. I desperately wanted to be in a project world where I could see my ideas getting transformed from paper to steel and concrete. To me, that was what engineering was all about. Everything else had to be subordinated to that.

When I earned my master's degree in 1986, the engineering market was still in a deep trough. The days of multiple offers awaiting graduates were a distant memory by then. As chemical engineers began exploring career options outside the petrochemical world, non-traditional employers began picking up the slack in the job market. As a result, some of my colleagues found interesting jobs in electronics manufacturing, pharmaceuticals, environmental engineering, or equipment manufacturing. As for me, my most promising prospects were with a food manufacturer, a pharmaceutical company, and a start-up equipment manufacturer/environmental consulting firm. It was the last prospect that really thrilled me: A super-small, no-frills company, with 25 employees in two locations in, of all places, Connecticut. (Nothing against Connecticut – it's just that it's not exactly a Mecca for chemical engineers.) The company manufactured ethylene oxide emissions control equipment *and* provided environmental consulting and permitting services to manufacturing companies.

It would not be an exaggeration to claim that I learned more in the 27 months I spent with a small company than I did in the rest of my career since then. There was, literally, no aspect of operations that I did not experience first-hand while working for my first employer. My assignments included stints in direct sales & marketing, engineering design, environmental permitting, technical consulting, contracting & procurement, equipment fabrication, equipment installation, start-up, testing, and customer support. Less glamorous, but no less exciting, were my foray into support functions such as human resource management, business stationary design, computer repair, and, at least in one instance, sanitary engineering (the unclogging of a bathroom sink). It was in this environment that I learned how easy it is to lose money and how hard it is to make a profit. Personal accountability and ownership took on a different meaning: What I personally did (or didn't do) could impact whether the bank was going to honor our paychecks at the end of the month or let them bounce like a house of cards. And yes, there was one month where we couldn't meet our payroll needs, and several employees had to dip into their personal savings and credit cards to make ends meet.

As you can surmise from the above, working for a small start-up company can be highly rewarding – both personally and professionally. These rewards, however, come at the expense of higher risk: The nature and volume of the work can be highly erratic, the growth opportunities can be severely limited, and the company can go bankrupt. If you can tolerate these risk factors, especially early in your career, the experience you gain from working with a small company are worth serious consideration.

Jumping Ship for the First Time

I mentioned earlier that leaving graduate school was one of the toughest decisions I made in my career. The toughest, however, was leaving my first employer. My risk tolerance was high, and I thoroughly enjoyed the work, the atmosphere, and the company of my coworkers. However, I realized that my growth opportunities were coming to a screeching halt. Fortunately, I came to this realization as the market for chemical engineers was warming up in the late 1980's. The oil bust of the mid 1980's was being replaced by an ethylene boom, and virtually all major E&C companies were actively looking for chemical engineers once again. The timing could not have been more perfect for me to enter the major projects world. I actively sought and then landed a role with a technology-based E&C company.

On January 16, 1989, I was one of 30 new-hires to start working for my new employer, a major E&C company headquartered in Houston. What I found ironic was that more new-hires started with me that day than the entire employee population of my previous employer. Gone were the days of writing my own proposals and purchase orders, and in were the days of strictly following procedures and operating manuals. I had to keep reminding myself that I didn't join an E&C firm because I enjoyed the bureaucracy of a large company. Rather, I was there to be part of an engineering team that will design and bring to life one of the most complex machines ever built by our profession: An ethylene plant. The bureaucracy and the procedures were simply means to an end.

The scale of a major engineering project is hard for an outsider to imagine: It takes upward of a thousand engineering work-hours to fully design and integrate a single piece of equipment into a chemical plant. Think back to your process design class in college where you may have prepared a preliminary process design for a simple distillation system. The system may have consisted of a distillation column with reboiler, condenser, reflux accumulator, and three pairs of pumps – 10 pieces of equipment in total. For these 10 pieces of equipment, an E&C company will spend approximately 10,000 engineering work-hours to develop a fully detailed specification package. Now imagine the scope of a major grass-roots refinery project, where the equipment count could be in the *thousands*. It is not unusual for a major E&C project to tie up *hundreds* of engineers and designers for *several years* from the start of the project to the finish. It is precisely this complexity that makes this type of work so exciting and challenging.

Chemical process engineering in project execution involves developing, among other things, heat & material balances, process equipment datasheets, process flow diagrams (PFDs), and Piping & Instrumentation Diagrams (P&IDs) for the plant. As important as we think our role is in project execution, chemical process engineering typically amounts to less than 15% of the total engineering effort on a project. The rest of the engineering work involves piping design, civil/structural engineering, control systems engineering, electrical engineering, machinery, equipment design, and environmental support. When viewed from this angle, the required oversight and bureaucracy to coordinate all of these activities in a consistent and efficient manner becomes clear. That is basically what a project engineer does on a major project.

The anatomy of an E&C project as presented above partially explains why the various engineering disciplines are highly compartmentalized. Process engineers may have a lead role in conducting process simulations and developing heat & material balances for the plant. However, they only play an advisory role for the other engineering disciplines. Similarly, equipment engineers may play a lead role in specifying engineered equipment (such as compressors, pumps, and exchangers), but they play only an advisory role in process design. The implications of this on career development are significant: An engineer at an E&C company will typically become a specialist in one discipline and may never get the opportunity to work in different technical disciplines. Coming from a background where I was used to doing many different things, this was a very hard pill to swallow. Fortunately, my supervisors were very supportive of my need to work in different fields: For many years, I was one of a handful of engineers who could claim "dual citizenship", one in process engineering and one in environmental engineering. The vast majority of my colleagues, especially those who grew up in the E&C world, were in "single citizenship" roles, and they didn't feel the need to venture too far from their Mother Ships. I bring this point up because there is an important attribute of the E&C world that is frequently overlooked: discipline longevity. There are no expectations for a newcomer to move from one discipline to another unless that happens to be what the individual wants. For example, a person who starts out as a process engineer could very well remain as a process engineer for his/her entire career without ever working in a different discipline. This is in stark contrast to the owner/operator world, where an individual is *expected* to move to different departments, disciplines, or groups as a broadening opportunity (more on that later).

With discipline longevity comes an extremely strong sense of belonging in the E&C industry. Members of one group or discipline could easily work together for decades on a myriad of projects without ever losing touch with each other. This results in tightly-knit networks and strong bonds among members of the same group or discipline. Nothing exemplified this better for me than by looking around the table at my farewell lunch: I had known and worked with each individual around the table either from the time I joined the company, or from the time they joined the company, whichever came later. There were no "outsiders" who came into the group through osmosis or diffusion from other groups; it was and will always be a *mass transfer limited operation.*

Discipline longevity also explains why most employees of E&C companies don't jump ship to owner-operators when the opportunity arises. In general, the E&C industry is considered to be more volatile than owner-operators when it comes to job security. Rarely is there a perfect balance between workload and headcount. As a result, the E&C industry goes through relatively frequent cycles of growth and contraction. The impact on the affected individuals could be devastating. With this omnipresent risk, one would expect that most employees of E&C companies would leap at the opportunity to join a more secure employer such as an owner-operator. A colleague of mine explained it best when he said the reason why he hadn't sought employment with one of his client firms is because "*I didn't want to have a different job in a different location every two years*". At the risk of over-generalization, the E&C world values depth and discipline focus whereas the owner-operator world values breadth and multi-discipline.

In no way am I implying that it's not possible for a chemical engineer to work in different *functions* at an E&C company. Personally, I had the fortune of being able to work in many different areas including process design & systems engineering, environmental engineering, project engineering, project management, and technology support. As my focus and interest shifted from purely technical roles to managerial roles, I took on the responsibility of leading a process technology group and then became a resource manager for the Project Definition department. As broad as these assignments sound, however, they were not that far from process engineering, which is what I was hired to do back in 1989. Even my office in 2006 was only a few feet away from where it had been 16 years earlier.

From E&C to an Owner-Operator

The E&C industry underwent a significant transformation in the early- to mid- 2000's. Much of the work shifted from North America to the Middle and Far East. The chemical industry was heading for a downturn, but the refining, LNG, and gas monetization sectors were on the rise. Suddenly, a $300 million project was no longer considered sizeable. Capital projects in the multi-billion dollar range were becoming more prevalent. The impact of this paradigm shift on the E&C industry was profound: Companies that were chemical technology-based began scrambling for engineering resources to staff the super-mega-projects for the refining and gas processing industries. For me, this shift meant that I had two options: Either align my career along a project execution pathway or look for a chemicals technology-based opportunity in a different industry. For various reasons, I opted for the latter.

Over the course of my career in E&C, I was fortunate to have developed close working relationships with a few of my clients. Through my contacts, I learned of an opportunity to manage a field technology group at a manufacturing complex. The technology component of the job certainly appealed to me, but I wasn't sure about the manufacturing support aspects. Clearly, it was a new area for me, and while I was happy to try it out, I didn't know if I would enjoy doing this line of work for a long period of time. Within an owner-operator organization, however, I knew that the range of career options was virtually limitless. That being the case, I was willing to take the chance that if manufacturing was not a good fit for me, I could seek a re-direction at some point in the future.

It should come as no surprise to anyone that operations or, more specifically, manufacturing support is about as different from engineering design as can be. Operation support engineers are like fire-fighters: They have to be tactical, comfortable working with incomplete data, able to think on their feet, naturally intuitive, thrive on adrenaline, and have an exclusive focus on getting broken things back on track. Most of their activities are short-lived, and long-term planning only goes as far as the next plant turnaround. Their time is not their own to manage: When the plant shuts down unexpectedly, it's all hands-on-deck regardless of the time of day or night, or what other activities had been previously scheduled or agreed-upon.

Process design engineers who move into plant support roles tend to get frustrated by the constantly shifting landscape of the production world. On the opposite end of the spectrum, production support engineers who move into the engineering design world tend to get frustrated with the level of detail and methodical approach required to develop a complete design package. Unlike the E&C industry where a process engineer doesn't venture too far from his/her discipline, a process engineer with an owner-operator has the luxury of being able to experience both ends of the spectrum. Rotation of developing engineers into operations support roles is not merely encouraged in an owner-operator organizations, it is a requirement for advancement in many cases. The idea, as I pointed out earlier, is for owner-operator engineers to develop breadth and become multi-disciplined as they advance in their careers. Many owner-operators, in fact, have highly structured programs that force new engineers to go through three or more rotations in different disciplines before allowing them to select their career paths. At my company, typical rotations include stints in capital projects, operations support, control systems engineering, and other disciplines. After the developing engineers complete these rotations, they are free to seek assignments in any field of their choice.

Although I had no intention of spending the rest of my career in operations support, I didn't realize how quickly the next assignment opportunity would emerge. A little over two years into my tenure, a managerial position with the R&D group sparked my interest. The nature of the work had much in common with what I had experienced in the E&C industry: In both settings, project time spans are typically of the order of a few years, and project execution is always methodical and sequential. In E&C, a project starts with a conceptual design and then proceeds through project definition, detailed engineering, procurement, and construction. In R&D, we typically describe a project using the "4D" model: Discover, Develop, Demonstrate, and Deploy. Neither setting affords you the luxury of skipping a step without assuming undue risk.

The 4D model concisely describes what a process development engineer experiences in an R&D environment. The process begins with a highly inquisitive individual discovering a new catalyst, a new molecule, or a new pathway to a product. A process evaluations engineer then conducts a preliminary study to determine if the novelty has technical and economic merit. If deemed feasible and strategic, the concept is tested in bench-scale rigs to establish an operational envelope and validate the technical feasibility of the critical steps. The experimental results from the bench-scale rigs are then used to develop a design for a suitably integrated pilot plant. The piloting step is, perhaps, the most important phase of an R&D project. Even though a project can be deemed infeasible at any phase of execution, it is the piloting phase that provides the most useful data to support a go/no-go decision. Although this phase is necessary, it is not always sufficient to assure successful commercialization of the concept. In many cases, a fully-integrated demonstration unit (semi-works scale) may be required for a variety of reasons. These may include verifying the scale-up approach, producing commercial quantities of a new product for customer testing, or providing assurance to investors that the project will, indeed, meet its financial objectives. If the demonstration unit meets all of its objectives, then the commercialization work can proceed.

Not every R&D project is successfully commercialized. Most projects, in fact, do not make it out of the "development funnel". This implies that development engineers may work for years on a myriad of projects without ever experiencing the commercialization phase in person. In an extreme case, a colleague who retired after a quarter of a century in this line of work casually mentioned at his retirement party that he had never worked on a project that was commercialized. For those of us who like to see their work in steel and concrete, this aspect of the R&D world is sure to bring some periodic disappointments. Conversely, an engineer who develops a project from concept all the way to commercialization will achieve a state of gratification that few in our profession will ever experience.

To MBA or Not to MBA

A highly "objective" colleague of mine once proclaimed: *Anything a mechanical engineer can do, a chemical engineer can do even better.* At the risk of plagiarizing my colleague's thoughts, I can add a highly objective corollary to his theorem: *Anything that an MBA can do, a holder of a graduate degree in engineering can do even better.* So why would a member of our profession willingly choose to dilute his/her analytical capability by pursuing a business degree?

If you cannot articulate *exactly* what it is that you will be doing with your MBA degree, you shouldn't even think about going to business school. The best advice on the subject can be found in a book titled "Business School Confidential" [1]. If you are thinking about going to business school, you owe it to yourself to read this book cover-to-cover before you make up your mind.

If you are pursuing an MBA to get into a different industry such as finance or management consulting, then good for you! That's where the degree would be worth your investment, and it will give you opportunities that would otherwise be unavailable to you as an engineer. Just make sure that you attend a top-rated school to get the best options after graduation.

If you want to stay with your current employer and believe that an MBA will give you different or better career options, think twice before taking the plunge. First, you need to make sure that the job you want does, in fact, require an MBA. Would your employer select you for this job because of your known track record or because of your new and shiny diploma? Second, think about whether an MBA is a good fit with the business model of your employer. The simpler the business model, the less valuable is an MBA degree to an employer. For instance, a mom-and-pop type business with one revenue stream and one cost stream has very little to optimize. On the other hand, a multi-national conglomerate with thousands of revenue streams will need hordes of MBAs to manage its complex operations. The E&C industry is certainly much closer to the simple end of the business model spectrum than, say, a major petrochemical company. This is not to say that an MBA wouldn't be valuable to an E&C company; Just that the opportunities for an MBA with an E&C company would be more limited than those in an owner-operator organization.

Finally, your interest in an MBA could very well be driven by a desire to learn the tools-of-the-trade rather than an interest in changing careers. In this case, you should pursue your degree on a part-time basis with a local school. This sounds a lot easier than it really is, especially if you have a family or an unpredictable work schedule. On the plus side, the distraction offered by part-time studies could be a welcomed change from the rigors of your day-to-day responsibilities at work.

I'll close this section by rephrasing my earlier tongue-in-cheek comment in a more serious manner: In my humble opinion, a graduate degree in science or engineering will enhance your analytical skills much more effectively than an MBA. You already have all the necessary skills to learn, *through self-study*, everything you need to know about finance, marketing, strategy, and organizational behavior. Have you ever seen an MBA learn engineering through self-study?

Closure

If your goal in reading this short paper was to identify the perfect job for yourself, I hope you're utterly disappointed by what you've learned from me! My goal was not to point you to Utopia but to share with you some of the realities of what it's like to work in E&C, manufacturing, or R&D. The best fit is clearly a personal choice. But more importantly, even a not-so-good fit can offer invaluable growth and learning opportunities - if you can keep an open mind and maintain a positive attitude.

Reference

Miller, R.H., and Koegler, K.F. Business School Confidential. St. Martin's Press, New York, 2003.

(69b) Process Evaluations & Manufacturing Support: From Concept to Commercial

Todd A. Whittemore, Team Leader - Process Evaluations , Shell International Exploration & Production Inc, Houston, TX

Through personal experience with projects ranging from minor debottlenecks to major grassroots petrochemical projects executed globally, Todd will share the importance of project development integrating product slate, process flowsheet, Licensor selection, partnering, option development, and techno-economic evaluation. Coupled with the initial selection of the overall project scope, Todd will discuss the importance of building a winning organization comprised of a delicately balanced blend of experience with strong local membership. Todd will use numerous real life examples highlighting both successful applications as well as major learning experiences to display the complexity involved to bring a mega-project from concept to commercial success.

Planning and Developing a Career in Process Development

Jonathan H. Worstell

Abstract

The first ten years of a chemical engineer's career are critical to the path that career takes, to the success achieved during that career, and to the happiness, fulfillment, and respect obtained from that career. This talk presents the questions which must be answered in the early years of a chemical engineering career. Some of those questions are:

- Am I best suited for a technical or management career?
- Do I need an MBA?
- How do I develop a technical career?
- How do I develop a management career?
- Should I get plant experience?
- Should I earn a graduate degree while working?
- How do I learn to be a leader?
- Where do I get leadership training?

These questions and many others will be discussed from the insight gathered during a thirty year career in Process Development.

Introduction

You can study a particular subject at many different levels. For example, consider biology. You can study biology at the molecular level; that is, inside a single cell. Molecular biology and biochemistry study life at this level. Identification of the structural features of a cell is done at this level also. This level is highly technical, being based on the laws of chemistry and physics. At this level, you study the forces impacting individual molecules and structures. The next level is cellular physiology --- the study of how the molecules and structural features of a cell interact to produce change and energy. This level investigates transformation --- it studies the flow of atoms and molecules into the cell, how the cell uses them to produce various molecules, and the flow of proteins and by-products from the cell. At this level, you study the forces impacting the coherent, unitized cell. The highest level studies how cell aggregates interact with each other; i.e., it studies living animals and plants. At this level, you study the natural history of animals and plants. Natural history includes the study of physiology, individual behavior; i.e., psychology, and social behavior; i.e., sociology. At this level, you study the forces impacting animal and plant populations.

A similar analogy can be made with the study of economics. You could study the economics of a single process or a single product in isolation from the national economy as whole. Such a study is called Process/Product Economics. Process/Product Economics is analogous to molecular biology and biochemistry. Process/Product Economics requires detailed technical knowledge of the process/product and applies the principles of chemistry and physics to the needs of humankind. The next level of economics studies how multiple processes/products forms individual businesses. This level is called Microeconomics and it is analogous to cellular physiology. Finally, you can study how a national economy functions. A national economy is an aggregate of businesses and business activities. This level is called Macroeconomics and is analogous to investigating the natural history of animals and plants.

Thus, the private sector economy has three levels: Process/Product Economics, Microeconomics, and Macroeconomics. Process/Product Economics is the lowest or simplest level. This level identifies trends impacting individual processes and individual products. Microeconomics is the next higher level --- it identifies trends impacting groups of manufacturing processes, product groups, and business centers. Macroeconomics is the highest level. Macroeconomics identifies trends impacting national economies and industrial groups. Figure 1 shows these generalized levels of the private sector economy.

The organization of a commercial entity; i.e., a company or corporation mirrors the organization of the private sector economy. The lowest level of a commercial organization is the Entry Level. Newly graduated engineers and scientists enter the commercial organization at this level. At this level, employees accept responsibility for learning the details of an assigned process or product. Thus, they become responsible for identifying trends impacting individual processes or products. This level corresponds to the Process or Product Economics level of the private sector economy. The next higher level in a commercial organization is the Senior Level. At this level, employees are responsible for identifying trends impacting business centers, groups of manufacturing processes, and product groups. This level is equivalent to the

Microeconomic level of the private sector economy. Organization Leadership comprises the highest level of a commercial organization. Organization Leadership is responsible for identifying trends in national economies and industrial groups. Thus, it mirrors the Macroeconomic level of the private sector economy. Figure 2 shows these generalized levels for a commercial organization.

Most newly graduated engineers and scientists have little knowledge of how the private sector economy is organized or of how a commercial entity is organized. However, such knowledge is imperative when planning and developing a career in the private sector.

Career Development Generalizations

After graduating from university, most of us work forty to forty-five years, then retire. Those years are filled with work and comprise our "career". Prior to those forty or forty-five years; i.e., during our educational years, we in general receive no information about career planning, career development, or even industrial or economic organization. Until you understand industrial and economic organization, it is difficult to plan your career and without a career plan, you cannot efficiently develop your career.

This laissez faire attitude starts in high school. When asked what we want to be, we answer an engineer or scientist, thus our career choice. However, beyond choosing to attend U of H or UT Austin or Texas A&M, we make no plans for developing our careers. When at university, career planning means following the required course sequence, thereby ensuring graduation ... career development means studying hard enough to pass each course. When interviewing for our first job, we tend not to consider career development: we are concerned with getting a job offer --- and our parents are just as concerned. Once hired, we fall into believing that the company will plan, manage, and develop our career, which may have been true in the past, but is no longer true today. Today, embracing that belief is perilous to career development. Today, you must plan, manage, and develop your career yourself. But then: who is better qualified to plan, manage, and develop your career --- you or someone in Human Resources?

Figure 2 provides guidance for planning and developing a career in a commercial organization in the private sector economy. Figure 2 shows that from the Entry Level, you proceed to the Senior Level and from there to Organization Leadership. Thus, a given career path can be traced as an arc inside such an organizational triangle. Figure 3 depicts several career paths. The length of an arc represents the relative time encompassed by a career. The flatter the arc, the faster a career develops.

To advance from the Entry Level to the Senior Level, you must prepare yourself to function at the Senior Level while performing the functions of the Entry Level. And, to advance from the Senior Level to Organization Leadership, you must prepare yourself to function as a Leader while performing the functions of the Senior Level. Therefore, you must formulate a plan that gives direction to your career. Unfortunately, most engineers and scientists do not put much thought into providing their career with direction: they "go with the flow", letting external forces set the direction and pace of their career. After forty or forty-five years, most finish their "career" vaguely dissatisfied, if not disappointed.

There are many paths from Entry to Organization Leadership. In a commercial organization utilizing technology to produce a product or a variety of products, you can proceed from Entry to Organization Leadership via a career in the Business Group, the Manufacturing Group, or the Technical Group. Management and Staff comprise each such Group, as shown in Figure 4. We are all familiar with the Business Group --- it is responsible for the supply and delivery of products to customers at the highest possible profit margin. The Manufacturing Group oversees the conversion of chemical and material inputs to product at the lowest possible manufacturing cost. The Technology Group support and improve current production processes as well as devise new processes for manufacturing product.

In small commercial organizations, many of the functions shown in Figure 4 overlap; thus, you may find yourself functioning in the Business Group and the Technology Group or some other combination of functions. Doing so broadens your organizational outlook. In medium to large commercial organizations, the functions shown in Figure 4 are separated. Therefore, you must spend time in each function to gain the necessarily broad insight into the organization required of Organization

Leadership. In general, you begin your employment in one part of the organization, then fill an assignment in a different part of the organization. Upon completion of this assignment, you may return to your original part of the organization and take a second assignment in the future or you may immediately take that second assignment immediately upon completion of the first assignment.

Newly graduated B.S. engineers are generally placed in the Manufacturing Group of an organization. Similarly, newly graduated B.S. scientists are placed in the quality assurance and environmental groups at manufacturing facilities. Most graduating M.S. and Ph.D. engineers and scientists are placed in the Technology Group of an organization. The technology Group may or may not be located at a manufacturing facility. If your Entry point is the Manufacturing Group, then you will want to obtain assignments in the Technology Group and the Business Group. If your Entry point is the Technology Group, then you will want to obtain assignments in the Manufacturing and Business Groups.

With the generalizations depicted in Figures 1 through 4, you can begin planning and developing your career.

Career Development
Career Planning

To develop a career plan, you must reserve a morning or afternoon so you can sit and cogitate undisturbed. Saturday is generally the best day for career planning: it is your time, not company time, thus you control it. If you are single, you can sit at your kitchen table and cogitate. If married, it is best to remove yourself to a quiet corner of the local public library or to your company's library or to a quiet table at your favorite coffee shop. Be sure to have an adequate supply of blank paper and a writing instrument so you can record your goals and document your plan for achieving those goals. With today's technology, the same can be done using a laptop computer.

To establish your goals, you must first decide where you want to be in twenty years. You have to know where you want to be in twenty years in order to know where you ought to be in ten years. And, you have to know where you want to be in ten years to know where you ought to be in five years. Thus, you will have three sets of goals: a five year set, a ten year set, and a twenty year set. You have to have all these goals because advancement from the Entry Level to the Senior Level requires you to prepare yourself for the Senior Level while fulfilling the functions of the Entry Level and advancement from the Senior Level to Organization Leadership requires you to prepare yourself to be a Leader while fulfilling the functions of the Senior Level.

Career Development
Technical or Business

Two career paths exist for engineers and scientists in most medium- and large-size companies. Those career paths are: technical-oriented and business-oriented. Both career paths provide growth for the individual engineer or scientist. The technical career path allows the engineer or scientist to develop technical skills in an ever-broadening economic context --- beginning at the Entry Level, progressing through the Senior Level, then concluding at Organization Leadership. The business career path allows the engineer or scientist to develop and apply business skills in the same ever-broadening economic context.

These career paths coincide upon hiring and during the first few years of a career, as shown in Figure 5. Thus, the first choice you face as a newly graduated engineer or scientist is which career path to follow. This decision should be made within the first three years of starting your career because subsequent career planning depends upon your choice. In fact, a career does not begin until you develop a plan. Such a plan provides direction and goals for your personal development. Without a career plan, you respond to the vagaries of your work and social environments. In other words, "you go with the flow" --- you do not direct the flow of your career.

Before making the choice between a technical-oriented career and a business-oriented career, you must analyze yourself. You must determine what type of person you are. Are you

- quiet or vocal?
- subdued or splashy?
- retiring or social?
- introverted or extroverted?

- passive or dominating?
- logical or intuitive?
- contemplative or abrupt?
- controlling or relaxed?

To assess your personality, take a blank sheet of paper and half it lengthwise with a vertical line. Label the left side "Strengths" and the right side "Weaknesses", then sit and review the above pairings and other personality traits you deem important. Record your results on your halved sheet of paper.

Remember, we humans tend to see what we want to see when viewing ourselves. Thus, do not be shy about asking others for input regarding this assessment. Ask your parents for input: they have only been observing, studying, and analyzing you for twenty-some years. Also, ask your friends for input for this assessment. When making such requests, remember that their "eye" is different from your "eye"; in most cases, they see you more clearly than you see yourself. Also, remember that your purpose is to match your personality with a career path: success, thus personal happiness, devolves from making the best match of personality and career. If you are extroverted, like meeting strangers, and enjoy interacting with all kinds of people, then you have the makings of a successful business person. On the other hand, if you are introverted and do not like meeting strangers or making small talk with others, and possess a modest understanding of mathematics, then you should consider a technically-oriented career.

With the above information in hand, you can make the most important decision of your career: whether to develop a technically-oriented career or a business-oriented career.

Career Development
Technology Management or Technical Resource Management

A career in Process Development means choosing the technically-oriented career. Note, however, that there are two aspects to Process Development: Technology Management and Technical Resource Management. Technology Management means you will be determining the current and future success of your employer. In other words, you will be improving current processes, thereby ensuring that product is manufactured at the lowest possible cost. You will also be identifying, developing, and implementing the next generation processes, thereby again ensuring that product is manufactured at the lowest possible cost. Technical Resource Management provides the technology staff with the resources necessary to support and improve current manufacturing processes and to develop new manufacturing processes.

If you do not like coaching people and if disciplining or firing people is distasteful to you, then you should consider pursuing a career in Technology Management. However, if you enjoy coaching people and if the thought of disciplining or firing people does not keep you awake at night, then you might consider a career in Technical Resource Management.

Invention, Innovation, and Sales

What Is Process Development?

There are many kinds of people inhabiting this world; three of them are Inventors, Innovators, and Sellers. An Inventor is one who comes upon, discovers a new concept, experimental fact, or way to assembly of equipment in a wholly different, beneficial manner. An Innovator is one who converts an invention into a commercial process or product and a Seller is one who induces, moves the public to buy the commercialized invention. For example, Benjamin Franklin proposed the concept of electricity and suggested that it flowed in a positive stream and a negative stream. Michael Faraday expanded Franklin's ideas and concepts in word pictures. Thomas Edison utilized Franklin's and Faraday's ideas and concepts to develop many instruments and machines to convert electricity into work. Those machines were manufactured and marketed by companies such as Western Electric and General Electric in the United States and Siemens in Europe. In this example, Franklin and Faraday are Inventors, Edison is an Innovator and the sales staffs of Western Electric and General Electric are Sellers.

In chemical terms, petroleum "cracking" provides a similar example of process development. Benjamin Silliman discovered in 1855 that the application of intense heat to crude petroleum decomposed; i.e., "cracked", the petroleum, thereby yielding volatile products. Between 1909 and 1913, William Burton of Standard Oil Company (Indiana developed a process utilizing Silliman's discovery and commercialized it at the Whiting Refining. Standard Oil Company (Indiana) then licensed the Burton process to other petroleum refiners. Silliman was an Inventor, Burton was an Innovator, and Standard Oil Company (Indiana) was Sellers. Thus, chemical process development is innovation.

Process Development
Job Description

Process development is innovation. It takes an invention and formulates, then develops a process that produces commercially viable product. Consider this example. The Greatly Expensive Chemical Company, a competitor of your employer, produces molecule C, a molecule much desired by the market, via the reaction

$$A + B \rightarrow C$$

Let us assume their process is complicated; i.e., expensive. And, due to the market's ravenous desire for C, your competitor gets their asking price, thereby making a tidy profit. This niche market needs competition. With considerable effort, the chemists at your commercial organization find that reactants X and Y yield C via the reaction

$$X + Y \rightarrow C$$

X and Y are not onerously expensive and you do a quick economic evaluation of an ideal process that suggests your employer can market C at a significantly reduced price, thereby gaining market share from Greatly Expensive Chemical Company. Thus, your employer decides to develop a process based on reactants X and Y.

The issues that you must resolve to produce C are

- TSCA registration of X and Y;
- inherent safety of X and Y;

- inherent safety of C;
- inherent safety of proposed process;
- pressure requirements;
- timeline using project management software;
- solvent usage;
- inherent safety of solvent;
- solvent purity;
- heat required or produced by the reaction;
- analytical methods;
- homogeneous catalyst;
- heterogeneous catalyst;
- reaction mechanism and reaction order;
- temperature dependence of reaction;
- by-product formation per order of reactant addition;
- temperature dependence of by-product formation;
- thermodynamic data;
- reaction engineering data;
- type of reactor --- batch, semi-batch, CSTR, bubble column, plug flow column;
- by-product formation per reactor type;
- by-product formation per extent of reaction;
- available laboratory reactors;
- available pilot plant reactors;
- process design options;
- separation of product from solvent, feeds;
- by-product (waste) management;
- solvent recycle;
- feed recycle;
- crystallization;
- drying;

- bulk handling;
- feed purity;
- solvent purity;
- product purity;
- PMN notice;
- MSDS;
- process engineering data;
- market sample;
- feed storage stability;
- solvent storage stability;
- product storage stability;
- semi-works availability --- use of contract manufacturing facility option;
- shipping for feed, solvent, and product --- tank wagon, rail tank car, barge.

This list is not comprehensive, rather it is suggestive. What this list indicates is the breadth of knowledge you need successfully this and other processes. Thus, Process Development is a multi-functional task.

Process Development
Required Knowledge Base

Pursuing a career in Process Development requires a thorough knowledge of the "first" principles of Physics and Chemistry and the application of those principles to everyday life and the real world. Thus, to maximize your career in Process Development, you need a sound grounding in Fluid Dynamics, Stage-wise Unit Operations (Absorption, Extraction, and Distillation), Heat Transfer, and Chemical Reactor Engineering, which means taking graduate-level courses. You can take these courses as a full-time student, thereby earning a Masters or Doctorate, then search for a Process Development position, or you can take these courses part-time, while working at a manufacturing facility or at a technology; i.e., R&D, facility. From a Process Development viewpoint, the most important point of these courses is the boundary layer concept --- the concept of a stagnant layer or film on both sides of a gas-liquid interface or a liquid-liquid interface or in the fluid phase adjacent to a solid surface, such as a catalyst surface or a heat transfer surface. This concept allows you to visualize the movement of mass from one phase to another phase.

You also need a solid foundation in Chemical Thermodynamics. First, Chemical Thermodynamics will tell you whether a process or process modification is possible. Second, in conjunction with Heat Transfer, it provides information about the potential safety hazards of a process or process modification.

The above courses provide the technical skills for developing a new process or for modifying a current process, but at some point, the process has to be transferred to a production facility. When that transfer begins, you will need a good understanding of how such a transfer impacts the environmental permit for the plant and the safe operation of the plant. It thus behooves you to take environmental courses and a process safety course to enhance your knowledge concerning these plant issues. Universities offer such courses, as do professional societies during national conferences. Remember; you do not have to be an expert in these areas, but you do need to know what such experts worry about and you need to know the vocabulary they use to express their worries.

Career Development
Twenty Year Goals --- Technical Fellow

Establish your twenty year goals first. Doing so provides direction to your career. Your twenty year goals also require you making your first career-oriented decision: to pursue a career in Technology Management or to pursue a career in Technical Resource Management. If you want to be in Organization Leadership, for example, become the Vice President of Technology or R&D, then you will tread the Technical Resource Management career path. If you want to be a Technical Fellow in twenty years, then you will tread the Technology Management career path.

Remember: in Organization Leadership, you identify trends that impact national economies and industrial groups. This level involves leadership --- functions through leadership. Managing and leading are different activities: managing involves transmitting direction from Organization Leadership; i.e., telling employees to follow that direction and monitoring employees to ensure that they are following the prescribed direction. Leading promulgates a vision and achieves influence through example. Leading is not telling, monitoring, supervising.

Both Technical Fellow and Corporate Leadership involve becoming a leader. Corporate Leadership envisions the commercial direction in which the company will evolve. Technical Fellows envision the technical direction in which the company will evolve

To lead, you must have a knowledge base broad enough to discern trends in national societies and national economies, then be able to integrate these trends to formulate a global picture of the future. Formulating such a picture requires an understanding of sociological and political trends. More importantly, it requires a broad understanding of scientific trends: which sciences and technologies are going to be economically controlling in the future. Thus, you function at the Macro level.

To be Technical Fellow requires developing leadership skills during your career, which means studying leaders and formulating a set of principles to govern your personal conduct. Reading biographies and autobiographies of acknowledged leaders will provide you examples of leadership and the skills required to be a leader. Such reading also provides you with "virtual" mentors.

A plethora of books on leadership exists --- which you need to read, particularly those entitled "the leadership principles of Churchill", "... of Washington", "... of Lincoln", "... of Grant", "... of Eisenhower", and so on.

To be a leader, you must formulate a set of core principles to govern your personal conduct and you must adhere to your principles. A leader is trusted by others; thus, to be a leader, you must develop a set of principles that will generate others' trust in you. All religions promulgate such a set of principles. In Christianity, that set of principles is the Ten Commandments. Reading philosophical ethics will also provide you material for developing a set of core principles. You cannot give "lip service" to your core principles: you cannot follow them on even days of the week and ignore them during the other days of the week. Others will trust you only if you adhere, faithfully, to your core principles.

Most Technical Fellows are recognized nationally as experts in a given process or product. Publishing refereed papers, technical articles, and/or patents provides such recognition. Presenting papers and leading discussions at professional society meetings, such as the AIChE and ACS, also provides national-level recognition.

Therefore, early in your career, you must

- formulate a set of core principles;
- read biographies and autobiographies of historically recognized leaders;
- read sociological and psychological literature;
- read macroeconomic literature;
- submit patent applications;
- write refereed papers and technical articles;
- participate in AIChE and/or ACS.

Career Development
Ten Year Goals --- Senior Process Engineer/Senior Research Chemist

Remember: at this level, the Senior Level, you identify trends that impact a business, process, or product. To identify trends that impact a business means interacting with a Business Group. Basically, this function identifies, justifies, prioritizes, and sells projects to a Business Group; i.e., gets a Business Group to authorize and fund a given project.

The best way to identify potential projects is to study the original DuPont Return on Investment (ROI) calculation, as shown in Figure 6. The ROI calculation shows the individual components controlling product costs. Potential projects exist in those cost numbers. Asking a manufacturing facility what equipment or process causes the most problems is another way to identify potential projects. Once a potential project has been identified, it must be justified; i.e., it must be analyzed economically. Such an analysis uses the tools you learned in your university Process Engineering course. After determining the economic value of a potential project, it is prioritized against existing potential projects. Net Present Value (NPV) is generally used for project prioritization. You will then present this prioritized list of projects to a Business Group biannually or annually. In other words, you will make a presentation to a Business Group that is actually a sales pitch. You will be selling projects to a Business Group. Such a presentation requires speaking skills. A project becomes "active" once a Business group authorizes funds for it. It then becomes your responsibility to manage the project and implement it commercially. Managing such a project means assembling and organizing a team of engineers and scientists.

To organize your project, use a project management software package, such as Microsoft Project. Outlining the work effort in Project shows the course and scope of the work effort and identifies the "rate limiting" steps embedded in the project, thereby allowing you to initiate their resolution before they become a crisis.

To assemble your project team, use concepts promulgated by Concurrent Engineering. If your project is for an existing manufacturing facility; i.e., process, include a plant process engineer on your team from its start. Doing so allows you to resolve plant issues as they arrive, rather than learning about them at the end of the effort, then restarting the project to resolve the identified issues. Including plant engineers in the initial project team accelerates development of the project and ensures and eases its implementation at an existing production facility. The same is true for Business Group personnel. Figure 7 shows a typical Process Development team.

Process development teams are generally large and decentralized with regard to location. Thus, keeping everybody informed requires effort on your part. Fortunately, e-mail exists today. Issue a weekly report to all concerned regarding progress per each item on your Microsoft Project timeline. Issuing a weekly report requires significant effort on your part, but it returns large dividends. First, everybody concerned can see the progress made weekly. Secondly, everybody can see who is becoming the "rate limiting" step and nobody likes such visibility; thus, everybody expends utmost effort on the project. Third, it is your the opportunity to write, thereby improving your ability to express yourself on paper or on an electronic monitor. Thus, early in your career, take as many writing courses as you can and read one or two how-to books on writing every year of your career.

Every project, at some time, requires face-to-face team meetings and teleconferences. With regard to such events, your goal is to keep the meeting or teleconference short and to the point. Invite only those necessary to make a decision and issue an agenda well ahead of convening the meeting. Identify on your agenda the issues requiring a decision. Ask participants to come ready to make the required decisions.

Therefore, to reach the Senior Level, you must learn to

- write quickly, clearly, and succinctly;
- make oral presentations;
- identify potential projects;
- justify potential projects;
- prioritize potential projects;
- sell potential projects;
- assemble multi-functional teams;
    - use Concurrent Engineering techniques;
- organize multi-functional teams;
    - use a project management software package;
- lead multi-functional teams.

You must also become an expert in one of the unit operations used in chemical processing and you must learn the scaling concepts for all the unit operations used in chemical processing.

Career Development
Five Year Goals

Your five year goals will be more detailed than your ten year goals. First, if you have not earned a Masters or Doctorate in Chemical Engineering, you should strongly consider earning a Masters or, at least, taking courses in Advanced Fluid Mechanics and Advanced Heat Transfer, as well as a course in Process Heat Transfer. These courses can be taken at a local engineering school where such courses are generally offered during evening hours. If you have a Doctorate in Physical Chemistry or Organic

Chemistry, then you should definitely consider earning a Masters in Chemical Engineering.

Second, if you are a plant engineer who wants to do Process Development, then you should inform management of your need for R&D experience. R&D experience will provide insight into the design of experiments necessary for Process Development. It will also provide insight about R&D's method for identifying, justifying, prioritizing, and executing projects. At the same time, you, as a plant engineer, will be able to educate the R&D staff about how plants view R&D projects, what concerns plants have about R&D projects, and what data and information plants require for approving and conducting plant tests, and how plants want the plant test organized, implemented, and executed.

If you are a process development R&D staff member without production plant experience, then you need to inform management that you desire a two-to-three year assignment at a manufacturing facility. Plant experience provides you insight into the issues that concern plant engineers about plant tests. Those issues are

- impact upon production;
- impact upon process safety;
- impact upon personnel safety;
- impact upon site environmental permit;
- equipment needs, temporary and permanent --- and the safe use thereof;
- data collection requirements;
- data analysis --- who analyzes the collected data;
- goals --- what constitutes success, who declares success or failure;
- commercialization of new process or altered process.

However, the main reason for plant engineers to get R&D experience and for R&D staff members to get manufacturing experience is to reduce the suspicion of each organization for the other. This suspicion exists because the mission of Technology Management is to wrought change while the mission of every manufacturing facility is not to change, is to continue producing product at maximum rate. By exchanging personnel, each organization gains insight into the purpose of the other organization.

Third, take the Professional Engineer's exam until you pass it. Passing it gives you *cache* with plant engineers and it provides you with more career opportunities in engineering. It also allows you to purchase liability insurance for your designs.

Fourth, learn and use a project management software program, such as Microsoft Project. Using such a program helps plan and organize a process development effort. It also identifies the "rate limiting" step of the project, thereby providing you a warning and allowing you to work the issue before it becomes a crisis. These programs also improve your productivity in general because they force you to plan and organize your work.

Fifth, hone your writing and presentation skills. At the Entry Level, you identify trends impacting individual processes or products. Generally, you will be asked to reduce the manufacturing cost of a process or product. Technical Resource Management will ask you to write progress reports as well as give presentations about you efforts to date. Use these opportunities to improve your writing and presentation skills. Read books and take courses of writing and presenting technical information.

Sixth, choose a unit operation or technical subject as you area of expertise. Your choice will depend upon which unit operations and technical subjects your employer uses, which might include some or all of the following:

- absorption;
- adsorption;
- corrosion;
- distillation;
- fluid dynamics;
- extraction;
- heat transfer;
- inspection/stress analysis;
- metallurgical analysis;
- process control/instrumentation;

- reaction engineering;
- thermodynamics.

Seventh, develop your Process Economics skills. If you did not take a Process Economics course at university, then enroll in an evening course at your local engineering school or take one of the short courses offered by various professional organizations. If you attended such a course during your undergraduate years, then consider taking an advanced course in Process Economics.

Eighth, during your early career years, you will participate on one or more multi-function teams. Observe how the team leader functions; analyze the leader's successes and failures and record your results in a "memo to file".

Ninth, join a professional organization, such as AIChE or ACS or both. Participate in the local organization in your area and in the national organization. These organizations provide you an outlet for technical articles as well as for making oral presentations. Becoming involved in their management provides you with an opportunity to develop leadership skills. Organize presentation sessions for their national conferences, chair such sessions, and join the leadership group of those subject areas which interest you. In particular, give attention to those sections covering Project Management.

Tenth, learn from your mistakes. Failed experiments teach us more than successful experiments do. Successful experiments simply confirm the validity of our theory and thinking. Failed experiments force us to reassess our assumptions and our theory. The same is true with respect to all endeavors. Endeavors evolving according to plan do not teach us much. Endeavors that go awry can educate you for future success --- if you learn from them. Thus, when an endeavor fails to meet expectation, take an afternoon to analyze what went wrong and why it went wrong. Be brutally honest with yourself. If you could have done better, then admit it. If you made errors in judgment, then admit it. Record your analysis in a recoverable form, either as an electronically stored document or as notes in a bound ledger; i.e., notebook. You will never improve until you admit that improvement is required. Your self-assessment provides you with data against which to measure future effort. Lastly, do not fear admitting mistakes to your supervisor. Freely admitting mistakes saves time and effort on everybody's part. It also shows that you assume responsibility for your actions --- a personality trait greatly admired by all supervisors.

Mentors and Mentees

Newly graduated engineers and scientists hear much today about the need for "finding" a mentor. They are told to find a mentor in order to reduce the time required to go from entry Level to Organization Leadership. However, mentees, those who are mentored, do not "find" mentors: mentors find mentees. Mentors are an organization's talent scouts. It is an unofficial role and a volunteer activity. A mentor is someone who keeps their eye open for the younger colleague demonstrating technical knowledge, personal responsibility, and self-initiated drive. Mentors know that such people require nurturing and promotion if the organization is to survive.

When newly graduated engineers and scientists hear that they need a mentor to accelerate their career, they assume the mentor has to be in Organization Leadership; i.e., has to be at least a Vice President; otherwise, how can a mentor accelerate their career to a similar lofty position. Unfortunately, newly graduated engineers and scientists rarely obtain the opportunity to present significant results; i.e., impress, Vice Presidents. Newly graduated engineers and scientists most often give presentations to Senior Level personnel. It is they whom newly graduated engineers and scientists impress. Thus, mentors are generally one level above a mentee. Senior Level personnel provide mentors for Entry Level personnel; Organization Leadership personnel provide mentors for Senior Level personnel.

Newly graduated engineers and scientists also assume that the mentor becomes responsible for their career development. Nothing could be further from the truth. Mentors provide promising individuals information about future technical and business trends, about organizational structure, and about how to navigate through that structure. Mentors do not take mentees by the hand and usher them into an Organization Leadership office. Mentors show mentees how to use information to produce the best possible work product.

Interestingly, while good mentors are "dime a dozen" --- good mentees are rare. The best documented career of a fully receptive, fully responding mentee is that of Dwight David (Ike) Eisenhower.[1] Ike was not the little known, lucky officer promoted to prominence as portrayed by many writers. Ike graduated from West Point in 1914 and was, by 1919, identified as a promising officer. In the early 1920's, Ike met Brigadier General Fox Connor, who became Ike's mentor. Figuratively, Connor poured information into an infinitely absorptive sponge, into Ike. And Ike responded by working hard to learn all that Connor could teach him. During the late 1920's, Connor passed Ike to the next level mentor: Army Chief of Staff Douglas MacArthur. MacArthur mentored Ike until mid-1939. Ike responded to MacArthur just as he had responded to Connor: he worked hard to learn all MacArthur could teach him. Throughout Ike's career, he was considered by those above him in rank to be the best, the outstanding officer of his age group and West Point class. He earned that reputation by working hard for his mentors, not by expecting those mentors to take his career in hand and walk him into a Leadership office. Within the Army, it was no surprise that Ike rose quickly through the ranks once war overtook America.

Leadership

Today, companies talk much about the need for leaders in order for commercial survival. Unfortunately, companies do little to train future leaders. At best, a company may issue a book about leadership, ask the recipients to read it, then schedule meetings to discuss the reading assignment. These meetings are generally led by someone who knows nothing about leadership. Thus, the exercise accomplishes little, except that the company can claim it has a leadership development program. The only organization that actively trains leaders from the day individuals join it is the American military. The American military understands the need for everybody to be a leader and it makes every attempt to provide leadership training.

Short of joining the military, being active in a professional organization provides an individual with the opportunity to learn and develop leadership skills. Such

[1] *Eisenhower Between the Wars: The Making of a General and Statesman*, M. Holland, Praeger Publishers, Westport, CT, 2001.

organizations provide opportunities to function at the Entry and Senior Levels outside a company. If leadership mistakes are made, as they generally are, they are made beyond company sight.

How do professional organizations provide leadership training? They provide leadership training by asking engineers and scientists to fill roles requiring others to accomplish given tasks --- without remuneration. Getting people to meet their goals without the "remuneration stick" requires leadership. Volunteers accomplish goals because they respect the person asking them to do so.

Professional organizations provide individuals the opportunity to learn what makes a leader. Thus, your participation in a professional organization benefits both you and your employer: you learn to be a leader; your employer gains a leader.

MBA: To Earn or Not to Earn

The question asked by every newly graduated engineer and scientist today is: "Should I earn an MBA?" Maybe it is not a question but rather a statement: "I plan to earn an MBA." During the early 1980's, few engineers or scientists had earned an MBA --- or even considered it necessary to do so. By the mid-1990's, most young engineers and scientists thought they needed an MBA to rise within a commercial organization. By the mid-2000's, almost all young engineers were pursuing an MBA in some kind of program: full-time, evenings, weekends, long-distance.

Three forces seem to be motivating young engineers and scientists to earn an MBA. First, realization during the 1990's that individuals on Wall Street and in corporate management were earning large salaries and exceptionally large year-end bonuses. Study of these individuals showed nearly all had an MBA. Lesson: if you want a chance at a hyper-inflated salary and bonus, then obtain an MBA. Second, from the end of WW II through the 1980's, when a young engineer or scientist looked up the corporate ladder, he or she saw each rung populated with Ph.D.'s. Thus, to move up that ladder, you considered earning a Ph.D. By the end of the 1990's, many rungs of the corporate ladder were populated by MBA's and by the end of the 2000's, almost all rungs were filled with MBA's. It is human nature to believe that if you have a Ph.D. it takes a Ph.D. to replace you. Any lesser degree devalues the effort you made to earn your Ph.D.

The same is true of an MBA: it takes an MBA to replace an MBA. If not, then why earn an MBA? Thus, the degrees populating a promotion ladder are self-propagating, self-fulfilling. This type of thinking is forcing all newly graduated engineers and scientists to consider earning an MBA once employed by a commercial organization. Third, the advent of the personal computer and spreadsheet software, such as Microsoft Excel, has made quantitative business calculations possible, as well as statistical analysis of large data sets. Much of an MBA program is teaching people to apply mathematics to business problems and to use the calculated results to formulate a decision. Prior to personal computers and spreadsheet software packages, such calculations were long, time consuming, and fraught with calculation errors. The result did not seem worth the effort, especially when applying statistical concepts to large data sets. Now, such calculations can be done in a fraction of a second --- so why not do them? In fact, today it is *de rigueur* to do them. Engineers and scientists are trained as undergraduates to use calculated results to make and support decisions. For them, applying the same thought processes to business-oriented problems is not a great step out of the proverbial conceptual box. However, such is not the case for art history, English literature, fine arts, history, political science, psychology, social work, sociology, or any number of other degrees that involve calculation-free thinking. For these latter degrees, earning an MBA is important because a new skill is learned --- calculation using spreadsheets. The degree of importance assigned to earning an MBA may not be as high for engineers and scientists since it does not necessarily provide them with an additional skill.

Conclusions

The purpose of this article is to help you plan, start, and develop a technical career in Process Development. The most important point of this article is: your career begins when you formulate a plan for its development. Prior to formulating such a plan, your personal development and career development are at the mercy of an array of forces dictating your response. With a career plan, you harness and direct those forces to shape your response to meet your personal development and career development needs.

As an engineer or scientist, your first choice when formulating a career plan is whether to remain an engineer or scientist or become a business person. This decision controls your career path to Organization Leadership in a commercial organization. Having made the choice to develop a technical career, you are immediately confronted with a second decision: to be in Technology Management or to move into Technical Resource Management.

Each of the above career paths will develop in response to the three basic levels of the private sector economy. Those levels are Process Economics, Microeconomics, and Macroeconomics. The Process Economics or Entry Level identifies trends impacting individual processes or products. This level requires a thorough understanding of engineering and scientific principles and their application to "real world" problems. In other words, this level stresses technical knowledge and technical competency. Your undergraduate education prepares you for this level. The Microeconomic or Senior Level identifies trends impacting business and process or product groups. Within a commercial organization, such as a corporation, it is a senior position. At this level, you are considered a technical expert in a given unit operation or engineering subject or scientific subject, such as catalyst chemistry, reaction mechanisms and kinetics, or thermodynamics. Also, at this level, you will be expected to identify, justify, prioritize, sell, and implement technical projects. Implementing projects means you need to know how to assemble a technical team and how to lead such a team. Therefore, to be qualified for the Senior Level, you must be able to

- write quickly, clearly, and succinctly;
- make oral presentations;
- identify potential projects;
- justify potential projects;
- prioritize potential projects;
- sell potential projects;
- assemble multi-functional teams;
    - use Concurrent Engineering techniques;
- organize multi-functional teams;

    - use a project management software package;
- lead multi-functional teams.

The highest level is the Macroeconomic or Organization Leadership. At this level, you identify trends impacting national economies and industrial groups. You must be able to discern trends in national societies and national economies at this level and integrate them into a global picture of the future. Formulating such a global picture requires an understanding of sociological and political trends and how such trends impact the direction the sciences and engineering technologies take. Such positions require a broad technical knowledge base, as well as a broad personal knowledge base. It also requires you to be a leader. Thus, you must develop leadership skills to function at this level of society. Thus, you must

- formulate a set of core principles;
- read biographies and autobiographies of historically recognized leaders;
- read sociological and psychological literature;
- read macroeconomic literature;
- submit patent applications;
- write refereed papers and technical articles;
- participate in AIChE and/or ACS.

Figure 1.
Generalized Organization of the Private Sector Economy.

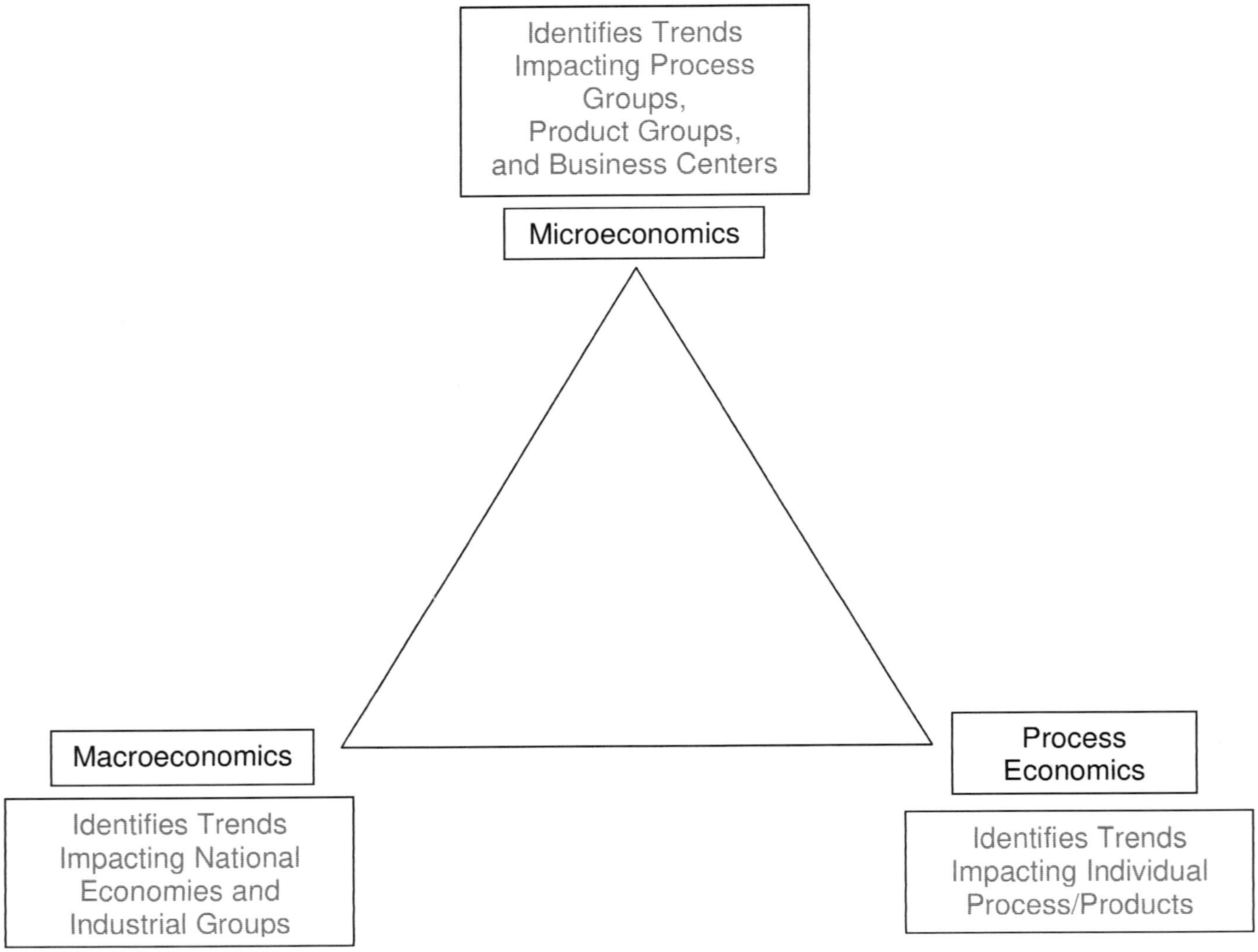

Figure 2.
General Structure of a Commercial Organization.

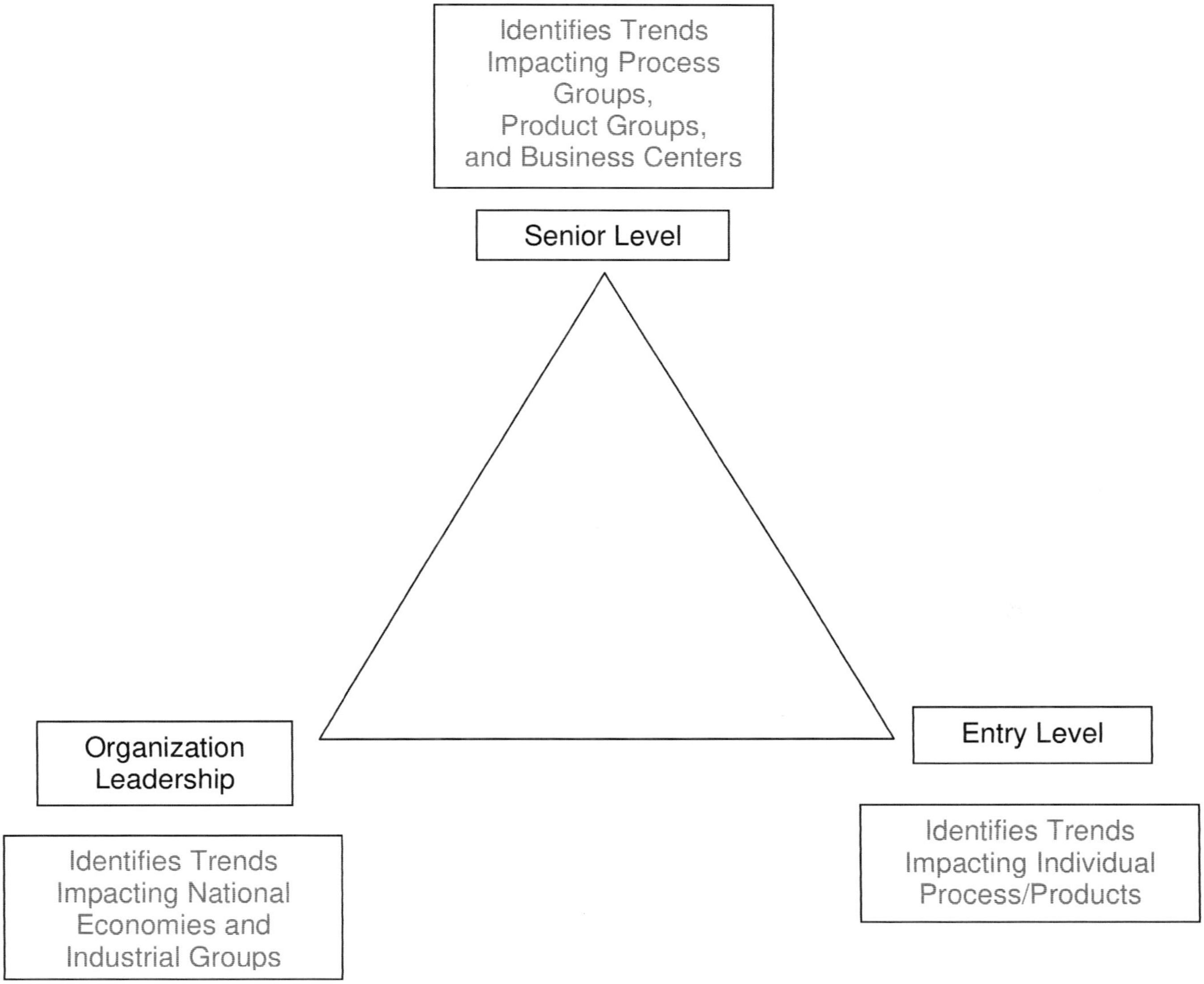

Figure 3.
Depiction of Various Career Paths.

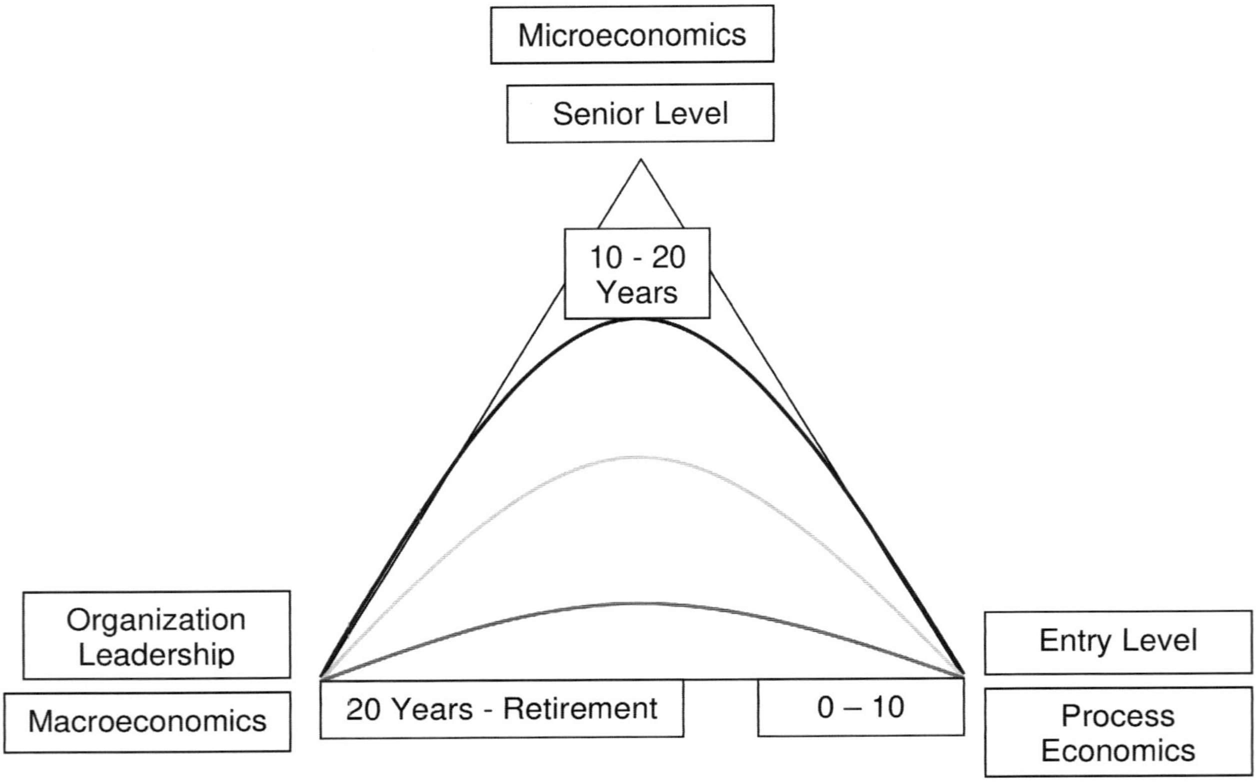

Figure 4.
Career Choices within Technical Companies.

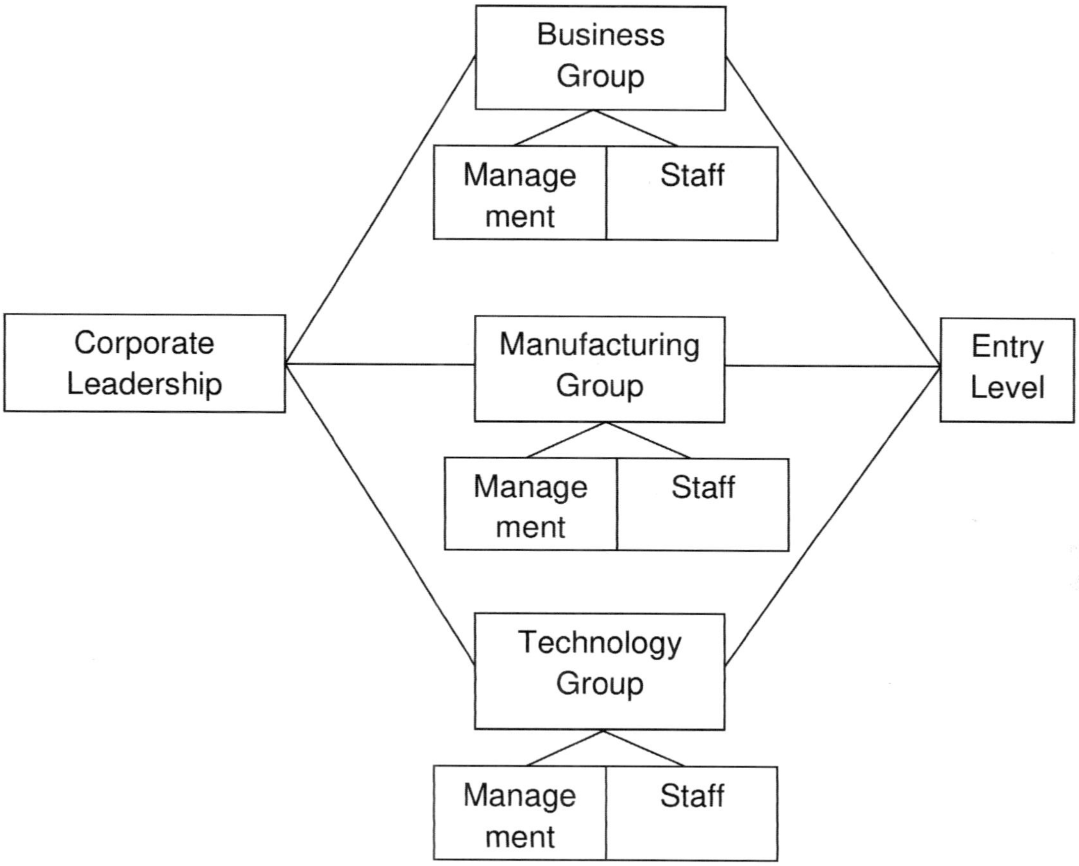

Figure 5.
Career Choices within Technical Companies.

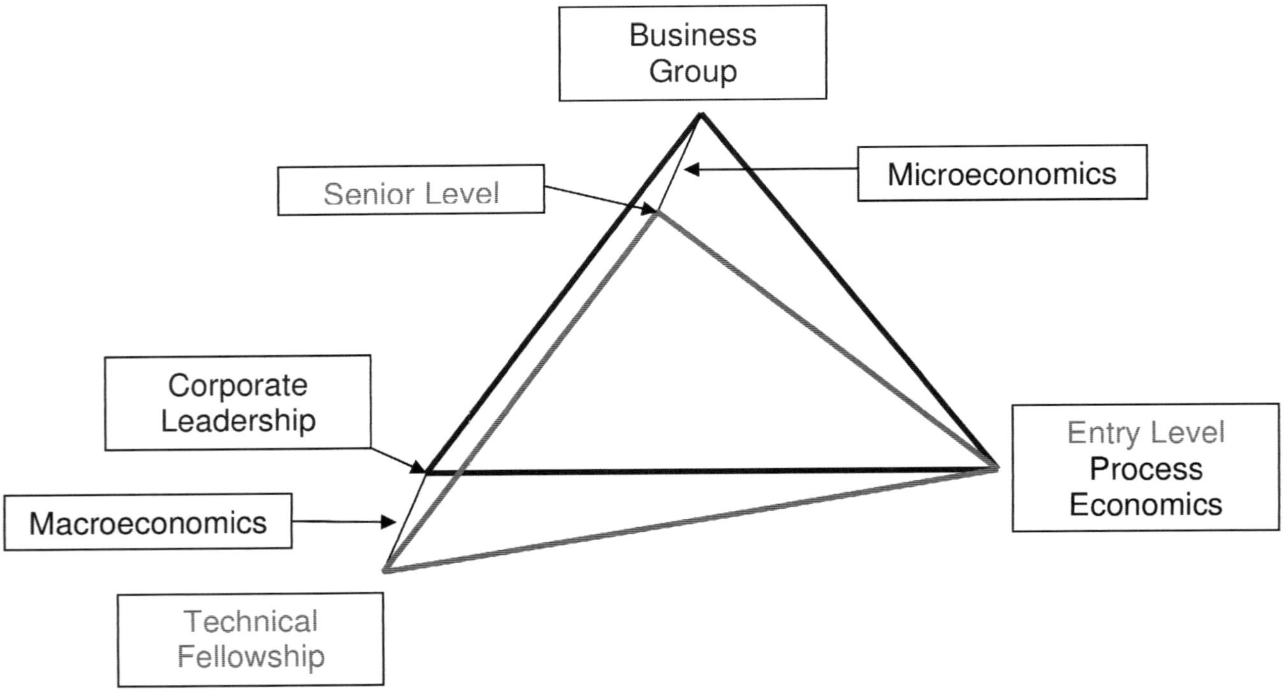

Figure 6.
Original DuPont Calculation of Return on Investment

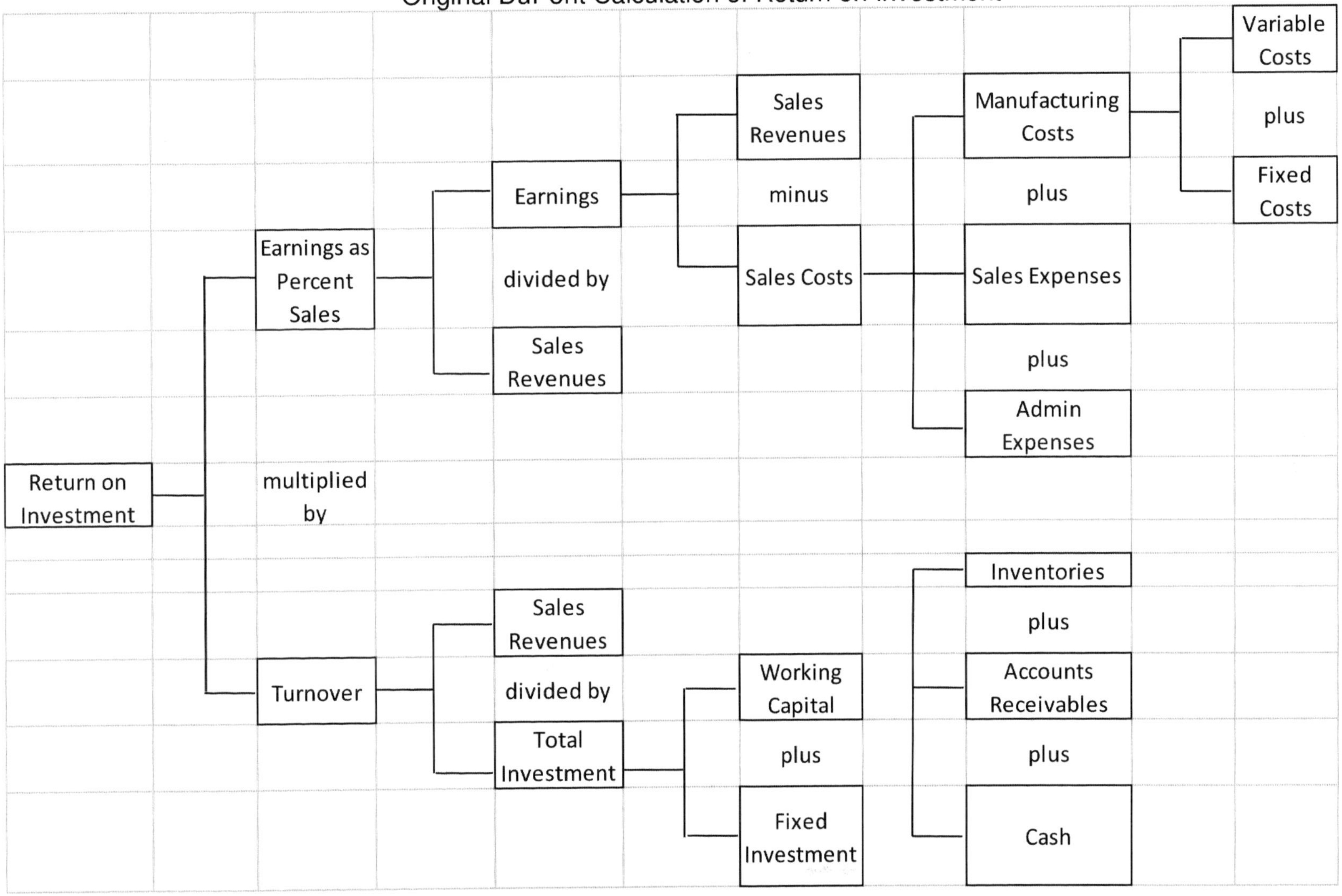

Figure 7.
A Typical Multi-functional Process Development Team

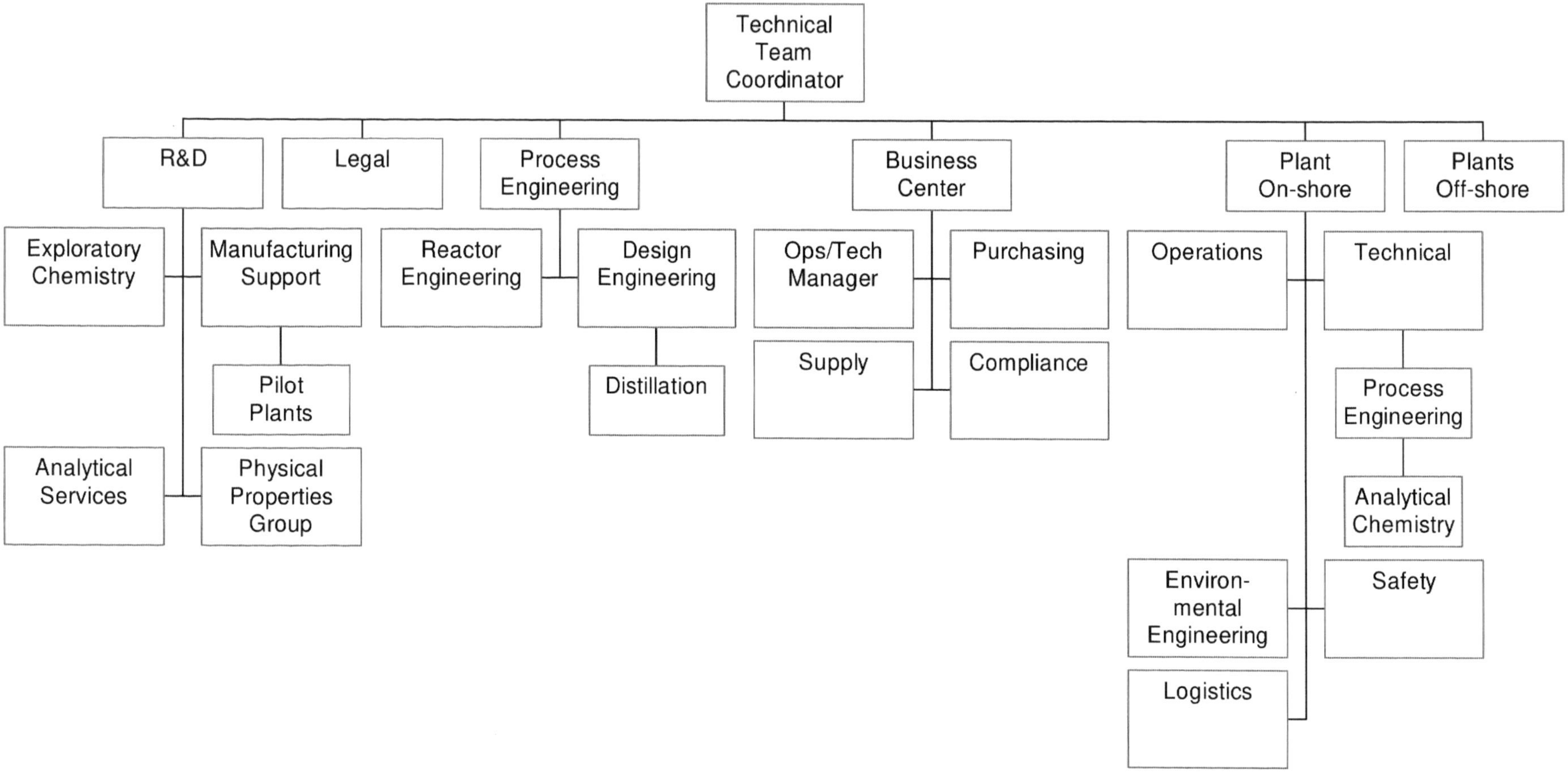

Process Development as an Industry Career Path

Joseph B. Powell, Ph.D.
Shell Chief Scientist – Chemical Engineering, Houston, TX

Introduction

While "process development" is not the subject of a separate course in the undergraduate or graduate-level chemical engineering curriculum, it plays a critical role in industry for the commercialization of new technology, and incremental improvement of existing technology. The following details activities, responsibilities, and opportunities associated with an industry career path in process development.

Process Development Discipline Areas

The relevant areas of process development were considered in setting up AIChE's Process Development Division 12 more than a decade ago, which grew from the "Pilot Plants" committee of the Engineering Group, with first chair dating back to 1984, to ultimately obtain full division status within AIChE in 1997. The initial areas commissioned for process development were [1]:

12A: Process Research and Innovation

12B: Pilot Plants

12C: Technology Transfer

12D: Manufacturing

While the division has grown to encompass Microprocess Engineering (12E), Process Intensification (12F), and Product Development (12G), the above four areas can be considered core to process development. Certainly, "pilot plants" is the most well-known sub-discipline under "process development". The construction and deployment of pilot facilities is the job description most commonly associated with this career path, and was indeed the founding basis for the process development division in AIChE. Scale-up and piloting of ideas generated from the "research department", for hand-off to "manufacturing", is one method by which process development can be implemented in industry. In companies where new product development is a core business opportunity, piloting of improved formulations can represent a full time opportunity. The various approaches to staffing and career paths for pilot plant support were documented in a benchmarking study by the Process Development Division's Area 12B [2], which identified however a broader definition of the role in many companies.

Stage Gates in Technology Development

There have been many names given to the various stages of technology development and deployment. Most recognizable for chemical engineering systems are:

- Research
- Development
- Demonstration
- Commercialization

Stages may be further defined in terms of readiness for deployment, including systems beyond those found in the chemical and process industries [3]. Processes developed by chemical engineers are typically sensitive to scale, and the above four categories are often used to depict the scale of study or implementation. Typically these steps entail bench scale research, one or more pilot plants (which are often the first integrated tests of process concept), followed by an optional demonstration unit to prove scale-ability or product viability, and finally, full commercial deployment. Rules of thumb for scale-up have been presented by Bisio and Kabel [4].

Process Development as the Integrating Discipline

While "development" is one of the stages along the path from idea to commercialization, it often plays an integrating role across all stages. Research may be driven for example by new chemistry, but in many cases the opportunity arises from a new catalyst, new process configuration, new unit operation such as a reactor or separation system design, or integration between unit operations as in process integration. Process development staff are therefore often highly involved in the research phase including idea generation, especially where engineering solutions are involved as part of the innovation.

"Development" leads to the initial prototype pilot units, but also an assessment of scale-up risk, from which the need for and sizing of demonstration units is decided. Core members of the process development team are therefore likely to be involved in demonstration projects, to reconcile scale-up behavior and process performance. New learnings also occur during commercialization [6], with six months typically required to achieve target rates and performance for a top quartile startup of new technology. The lead process development engineer and other core team members are also often involved with pre-commissioning, startup, and initial commercial operation, to effect "technology transfer" via extensive interaction with manufacturing.

Key chemical engineering sub-disciplines involved in these activities are the ability to do process synthesis for selection among competing options and unit operations, and economic analysis [5]. Chemical engineering training is perhaps unique among the engineering disciplines in emphasis on systems design and economics, and thus provides the basis for deciding between options, which can only occur after thorough grounding in business principles and drivers. Since all steps needed for commercial deployment comprise the process design, the process development engineer must become expert in virtually all aspects of chemical engineering, from chemistry and catalysis, through reactor design, optimization and selection among plausible separation options, as well as process control, metallurgy, safety and environmental stewardship. Interactions with demonstration or commercial project teams, engineering and construction partners, and ultimately a manufacturing location plus its business sponsors provide insight and training into the needs, activities, and work processes of the broader organization.

Along the complex path from idea to commercialization, a great number of skills are required, and typically gaps or "white spaces" arise to warrant learning of a new skill or core competency in fundamentals of chemical engineering, or other business, scientific, or management systems skills. Process development can thus be a visible springboard to a business career, a manufacturing location career, or a continuing career in R&D with focus on innovation, maintaining core technical skills, and knowledge of best practices in technology de-risking and development. Just as the AIChE Process Development Division has grown since 1984, opportunities in career development extend well beyond "pilot plants" for a career path in this discipline.

References

1. AIChE Process Development Division: http://www.pd-aiche.com/structure.html.
2. Bob Duggal (Afton), David Edwards and Sean Murray (Zeton), Daniel Pintar (UOP), Joseph Powell Shell), Pilot Plant Benchmarking Study Results. http://www.pd-aiche.com/area12b/pp_benchmarking_results.html
3. Mankins, John C. (6 April 1995). "*Technology Readiness Levels: A White Paper*". NASA, Office of Space Access and Technology, Advanced Concepts Office. http://www.hq.nasa.gov/office/codeq/trl/trl.pdf.
4. Bisio and R. L. Kabel, *Scaleup of Chemical Processes*, John Wiley & Sons, NY (1985).
5. J. B. Powell, *Economics of Pilot and Development Units for De-Risking of Commercial Processes,* **AIChE Spring National Meeting Conference Proceedings, San Antonio**, TX, *March 21-25, 2010.*
6. J. B. Powell and J. N. Chheda, *Assessing Potential Value of R&D Opportunities Via Explicit Assumption Vs. Implicit Experience- or Learning-Curve Approaches,* **AIChE Annual Meeting Conference Proceedings 2011**, October 16-21, Minneapolis Convention Center, Minneapolis, MN.

() Musings on Management

Frank van Lier, The Lubrizol Corporation, Wickliffe, OH

It really is all about the people!

Hiring good people.

Retaining them.

Adjusting for different personal styles, cultures, encouraging diversity, embracing different perspectives.

Motivating people to work together in successful global teams.

Achieving top notch results that are better than any individual can achieve alone.

Maintaining the focus.

Getting the results.

Management is all about the excitement managers gets from having their people and teams grow and succeed.

(77a) Cultural and Technological Forces That Are Changing the Knowledge Management Game

Cindy Hubert, APQC, Houston, TX

With constantly changing technologies and work force demographics, developing innovative approaches to help your organization's critical knowledge flow is becoming increasingly essential to business advancement. Hubert will discuss these key points, including the steps to building and maturing knowledge management capabilities, generational concerns in knowledge transfer and retention, social networking on the Web, and other significant questions raised by an evolving business landscape.

() What Kind of Manager Am I?

Terrence Robak, T Robak Associates LLC, South Plainfield, NJ

Terry Robak is a facilitator, coach, trainer and speaker with 25 years experience in the organization and leadership development arenas. Experienced college-level instructor with line operations management experience and a proven ability to design and deliver learning to students. A Certified Professional Co-Active Coach that blends coaching and facilitation skills to provide maximum results from groups. A master trainer in leadership and communication skills that integrates personal business experience with theory that is described as: "challenging, energetic, caring and very professional."

Best Practices in the Design and Fabrication of Pilot and Demonstration Plants

David Beckman, President and CEO, Zeton Inc., Burlington, ON, Canada
Adam Whalley, Business Development Manager, Zeton Inc., Burlington, ON, Canada
Presentation ID# 246699

AIChE
2012 Spring Meeting
April 1-5, 2012
Hilton Americas-Houston and George Brown Convention Center
Houston, TX

As a specialty designer and fabricator of modular pilot plants, one of the first questions asked of Zeton in the early stages of a project is 'how can we speed up our project, and get the pilot plant delivered in less time?' The engineers asking this question are experienced professionals that understand project management. They already have a firm grasp on fundamental tactics that all good project managers know. Project Managers understand schedules can be compressed by expediting suppliers, overlapping work tasks, compromising on materials of construction, working overtime, shift work and relaxing specifications.

A step change in project execution speed cannot be achieved by applying more of the same type of tactics. Although these common tactics are important and all good project managers should be using them to speed up design and fabrication, they soon reach a point of diminishing returns. For instance, five mechanical technicians might fit comfortably working within one area of a module, and fabrication can be sped up if that number is increased to 7, but efficiency starts to decrease when the skid becomes crowded. Adding more personnel to that area will lead to further congestion, safety problems, and possibly an even longer schedule.

Increasing project execution speed is possible, but only through setting up the project at the beginning with a different strategy. The traditional design and fabrication set up breaks the job into pieces: full detailed design and costing first, followed by procurement and fabrication. Significant schedule reduction can be achieved doing these activities in parallel. The philosophy that leads to a step change in reducing schedules is to get access to pilot plant metrics, and locate the design and fabrication team at the same facility, which allows an Integrated Approach. A secondary, but notable benefit of the latter point is that safety can be improved by locating the design and fabrication team away from the final installation location.

The chart in Figure 1 illustrates the differences between a traditional engineering and fabrication approach and the Integrated Approach of a specialty fabricator. In an article written for Chemical Engineering magazine, Rich Palluzi sums up the main reason for the differences in a few words:

> "rarely is there enough information available for pilot plant projects to make an intelligent estimate of these costs, unless one has a great deal of historical data." P. 52

Palluzi, R. (2004). Pilot plant cost estimating: Make intelligent use of contingency. *Chemical Engineering, Volume (Issue),* 52 – 56.

Mr. Palluzi was discussing contingency in this particular article, however the same principle applies to working out the costs and schedule associated with detailed design, procurement, fabrication and factory testing of a pilot plant. Companies that are untested in the design and fabrication of modular pilot plants, or even modular commercial plants, do not have the metrics available to estimate costs with limited data. For that reason, a company lacking this experience and data is limited to completing full detailed design before entering a bidding phase for fabrication, as illustrated in the Traditional approach in Figure 1.

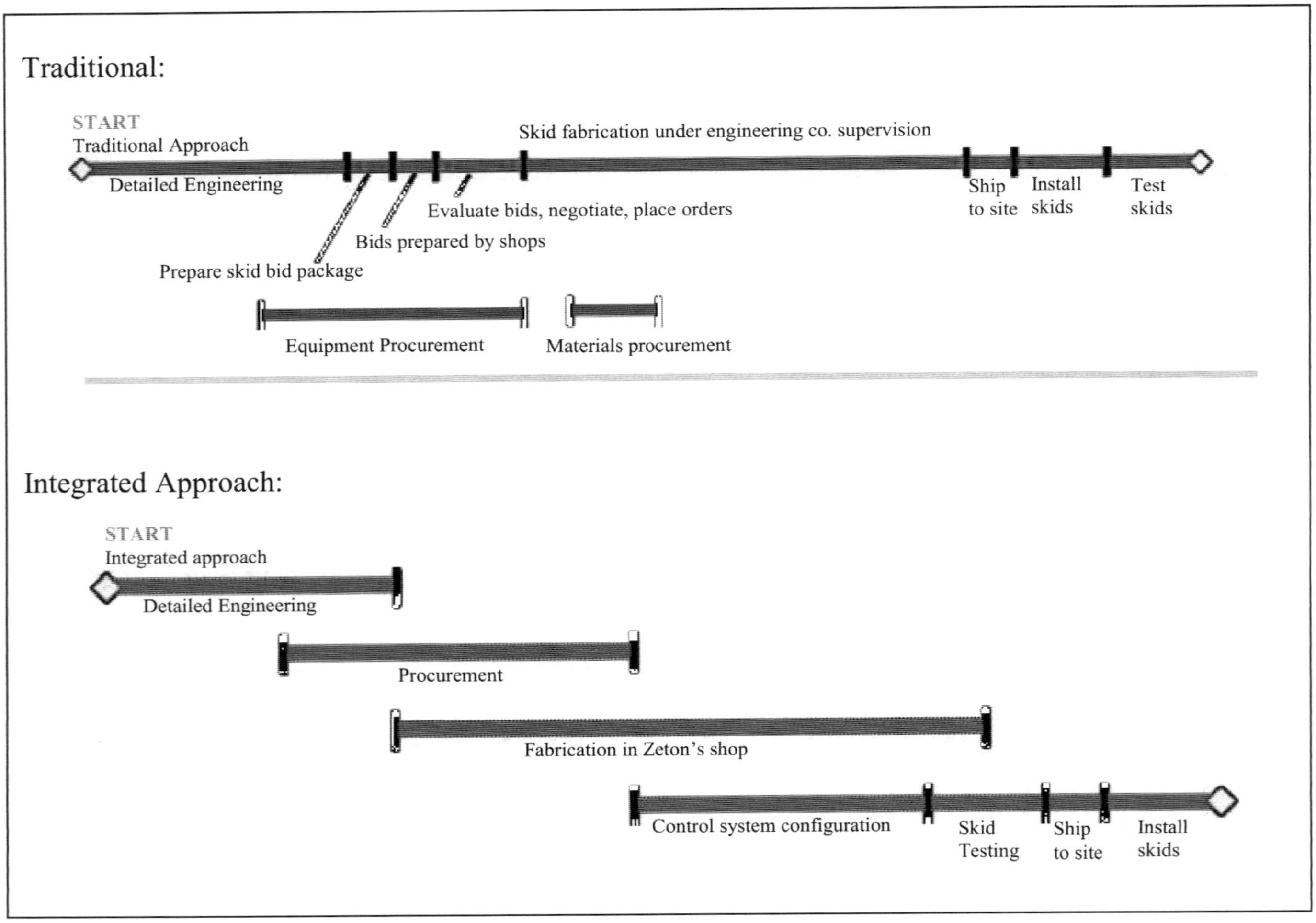

Figure 1: Traditional versus integrated design and fabrication approach.

On the other hand, specialty modular pilot plant companies have collected hundreds of data points. Once prices have been fixed in three main categories, major equipment, instrumentation, and hand valves, and some other limited information collected, the specialty company can work out an accurate price that can form the basis for a lump sum contract. Access to pilot plant metrics impact schedule directly by limiting the amount of detailed engineering that needs to be completed before the next phase.

Although all detailed engineering needs to be completed at some point throughout the project, only about one-half of it needs to be completed at this early stage. Depending on the size of the project, the first phase called Basic Engineering can last as little as 8 weeks or as long as 7 months. In other words, reducing the amount of time required by one-half cuts this part of schedule down by 8 – 28 weeks depending on the size of the project.

Of the remaining amount of detailed engineering work to be done, about one-half can be overlapped with procurement and fabrication as illustrated in Integrated approach in Figure 1. As mentioned above 8 – 28 weeks can be saved in Basic Engineering by having access to a large volume of metrics collected over hundreds of jobs. Nevertheless the time savings gained during the first stage of detailed design does not turn into an equal savings on the overall schedule. The final decrease will be about ¼ of total detailed design. So about 4 weeks on a small project and 14 weeks on a large project.

The second step change in schedule is realized by locating an experienced pilot plant design and fabrication team under the same roof. The most important part of this philosophy is that communication is greatly improved. As can be seen from Figure 1, part of detailed design is overlapped with fabrication, which is done by providing 'just-in-time-design' to the fabrication shop floor.

Close involvement of the engineer throughout fabrication ensures fabrication continues unimpeded. The point is not about cracking the whip to ensure all mechanical technicians are working hard. Rather it is for the engineer to act as a resource to the fabrication team, so obstacles can be removed from their path as the project proceeds. If any questions or uncertainty arises they are immediately answered and resolved The end result: less standing around waiting for orders and issues to be resolved; constant work flow and reduced schedule; better quality work – less miscommunication.

Another distinction between the traditional approach and the integrated approach is the potential for rework. During the traditional approach, the engineer might make a trip to the fabrication site once per week. The mechanical technicians in this case are not be specialized. This combination infrequent communication and non-specialized labor creates a high probability for incorrect installation and rework.

The mistake is multiplied further when it happens shortly after the engineers visit to the fabrication shop. If weekly visits are the norm, the next inspection of the work will happen the following week. A lot of work can take place in one week, and much of it will be dependent on the area where the mistake was made. Undoing this work to fix the error adds schedule risk. The reason traditional projects prepare a full and complete detailed design package prior to fabrication is to eliminate the need for constant engineering supervision. The drawback is that schedule suffers.

As mentioned earlier, with the integrated approach engineers who understand process concerns are on the shop floor throughout the entire project duration, and available to answer questions. This availability not only removes informational obstacles for the fabrication team, it also reduces assumptions and mistakes. Mechanical and electrical technicians are more likely to install parts the right way when they have fast and accurate answers. Errors are less likely to occur in the first place reducing schedule risk.

Despite the project manager's best efforts, mistakes happen regardless of the quality of the fabrication plan and schedule. The difference between the two systems is that when errors are made during the integrated approach, they are caught and fixed immediately. Preventing rework that would happen as a result of missed mistakes saves schedule further, and can be the difference between meeting and missing the deadline.

Planning for the deadline involves gaining access to metrics for building pilot plants. Specialty companies using an integrated approach have designed built hundreds of pilot plants. Through this experience, large databases can be developed that collect and track the mechanical and electrical technician hours required to complete tasks for various types of pilot plants. The final hours predicted to complete fabrication are based on a crew that specializes in building pilot plants. In many cases, the technicians have worked on as many as 100 pilot plants. When contrasted with a traditional fabricator that does not have experience with modular pilot plants, it is reasonable to expect more hours to be required from their fabrication team. Quoted schedules from specialty pilot plant companies can be much tighter based the specific skills of the fabrication team.

Further to speeding up fabrication, the integrated approach also makes it possible for the technology developer to complete other important activities in parallel. Preparing the site to receive the pilot plant modules prior to their arrival is very realistic. Also, while the plant is being built, research and development efforts continue in parallel, often leading to changes in the process technology that must be incorporated into the plant. This requires a flexible, disciplined and integrated approach between the specialty pilot plant company and the technology developer, such that the project continues moving towards completion.

The last step that impacts schedule is testing the entire integrated pilot plant. The most important date for schedule driven projects is the date the plant starts producing product. Clients sometimes request the plant be installed at site prior to testing, but this is not recommended. It might make sense from the perspective of a funding cut off date, or some other public relations announcement, but it rarely speeds up commissioning. In some cases, it is difficult to complete a full test. For instance if 100 ft towers are built in two pieces, they cannot be stacked and tested in the same orientation as the installation location. In other cases, the entire plant can be connected and run in the proper orientation. In either case, a factory acceptance test with simulating fluids should always be completed in the shop to the maximum extent possible to reduce the overall time to start up.

At Zeton, the engineers that design the units are always involved with the tests, and they are mainly looking for proper operation of rotating equipment and instrumentation. When a part of the plant is found to be operating below the specified limits, it can be fixed faster in the shop than it can be on site. The shop is equipped with the right tools such as overhead cranes, and a clean, well-light, work space. Testing can be completed quicker, because technicians are experienced from their work with many other pilot plants. Vendors local to the fabrication facility can fix their products immediately. In other cases where the vendor is not local, malfunctioning equipment can be removed and returned to the vendor while the plant is being trucked to site. There is a good chance this equipment can be replaced and installed prior to commissioning. Although the plant arrives later at site, testing at the fabrication facility always reduces the overall schedule by saving time during commissioning.

When all of this work is completed in the shop of a specialty pilot plant company, overall project safety is improved. Roland Meissner III commented in his article 'Modular versus conventional construction: Choosing the right path'

> "offsite prefabrication avoids plant hazards such as the performance of hot work near adjacent operational units, and can overcome access restriction that arise in an operating process plant" p. 49.

Meissner III, R. E. (2003). Modular versus conventional construction: Choosing the right path. *Chemical Engineering, September 2003, Volume, (Issue)*, 46 – 52.

Also, building the plant off-site in a controlled indoor environment improves safety, especially on larger projects. Projects stick-built outdoors must contend with rain and snow, and these slip hazards are avoided in a controlled environment. Overhead cranes in the fabricators shop provide an element of control that improves safety over stick built projects.

The answer to the question "how can I get my pilot plant faster" is to find a way to gain access to pilot plant metrics, and ensure a specialized design and fabrication team is located at the same facility, allowing design and fabrication to proceed in parallel. Without a large database of pilot plant design and construction metrics, detailed design needs to be completed in series with fabrication. With the metrics, 4 to 14 weeks can be cut out of the schedule as a result of overlapping about ¼ of the detailed design tasks with fabrication. The second step of locating the team under the same roof is a prerequisite to overlapping detailed design with fabrication. Just-in-time-design will not work any other way, because communication needs to be daily and in person. Reduced errors and less rework are inherent advantages of this structure, which have a sizeable impact on improving schedule, or at least hitting the target completion date. When design and fabrication is outsourced to a specialty pilot plant company, research and site infrastructure work can proceed in parallel, with an added benefit that safety can be improved.

ExxonMobil
Research and Engineering

Best Practices in Heat Tracing Pilot Plants

Richard Palluzi
ExxonMobil Research & Engineering Company
Clinton, NJ
April 4, 2012

Introduction

- Pilot plants are often required to operate at elevated temperatures
- Keeping all piping and components heated can be challenging
- Some industrial methods are too difficult or costly on the pilot plant scale
  - Wrapping is rarely possible in smaller size tubing
  - Many traces are larger than the tubing leading to poor contact area
  - Piping complexity leads to proportionally more heat sinks
    - + Supports
    - + Valves
    - + Branches

Major Types of Heat Tracing

- Electric
  - Slightly more expensive than other methods
  - Easier to install, maintain, modify, and reuse
  - Types:
    + Self Limiting
    + Constant Wattage
    + Tubular Resistance Heaters
- Steam
- Hot oil
- Heated enclosures

3

ExxonMobil
Research and Engineering

BEST PRACTICES

4

ExxonMobil
Research and Engineering

Self Limiting

- Make sure that the heat output is at the desired operating temperature
  - Tracing output is a function of operating temperature
  - Adding tracing may not help
  - Recognize that the maximum temperature they can reach is usually significantly less than the published values
- Start up requires significantly more power
  - Restricts lengths per breaker often requiring multiple zones for larger applications
  - Starting at a lower temperature may not be possible!

Self Limiting

- Experience has shown that significant conservatism is usually required
  - Poor heat transfer from tracing to tubing due to geometry
  - Plan on 20-30% extra power
- Wrap generously
  - Include supports, dead legs, gauges, instruments, etc.
- Heat transfer cement and heat transfer tapes rarely very effective
  - Help but rarely cost effective vs. simply using a higher wattage tape

Constant Wattage

- Quality varies widely from good to poor
  - Lower cost laboratory supply house models are often non-uniform and have a low service life
  - Heavier duty (and more expensive) commercial versions often last much longer
  - In house testing is recommended
- Connected components can get so hot as to create a burn hazard
  - Standoffs or insulation barriers are usually required
- Installation in an electrically classified area is very difficult
  - Heater wires almost always exceed autoignition temperatures no matter how low the control temperature
  - Any leaks will have to pass an ignition source

7

ExxonMobil
Research and Engineering

Potential Solutions

- Purchasing higher voltage tape and running it at lower voltage (e.g. 120 VAC on a 240 VAC tape)
  - Depends on numerous factors so testing before use is recommended
  - Be careful that additional insulation will not inadvertently lead to higher temperatures
- Limiting the voltage (variable transformer) or amperage (phase angle fired SCR)
  - Many SCR's cannot limit either and the heater wire will reach operating temperatures but for a shorter time which is not safe
  - Again testing is recommended
- Putting a thermocouple on the heater wire is desirable but almost impossible in practice

8

ExxonMobil
Research and Engineering

Constant Wattage

- Poor wrapping techniques can create hazards
  - Crossing or doubling can lead to a fire
  - Shorting to or even through piping has occurred
  - Easily confused with wrap on insulation when removed and replaced
  - Need to leave clearly defined spacing between wraps

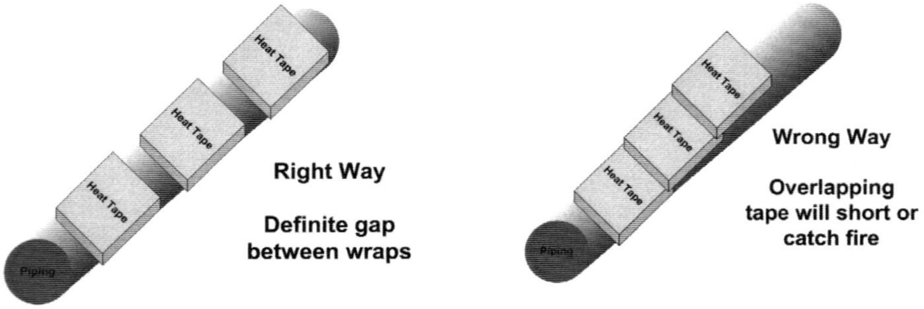

Constant Wattage

- Tapes are both non-uniform and often high heat capacity which results in significant temperature variation
  - Suggests that use should be limited to cases where higher temperature variations can be accepted
  - Some trial and error testing may help minimize
    - + Rewrapping with more spacing between wraps
    - + Adding or removing insulation in specific areas
    - + Time consuming and often of limited effectiveness
- Burn hazards are very common
  - Often difficult to identify in advance
  - Usually found only after a burn is received
  - Extreme care in initial testing and operation is required
  - Thermal imaging devices can be very helpful in identifying
- Heat transfer cement and heat transfer tapes rarely very effective

Tubular Resistance Heaters

- Connected components can get so hot as to create a burn hazard
  - Standoffs or insulation barriers are usually required
- Laying the tubular heater against the piping is recommended
  - Wrapping can break the heater if the bend radius is too close
  - Reuse (wrap-unwrap-rewrap) usually leads to fatigue failure
- Numerous, closely space ties are recommended
  - Heaters tend to pull away when hot, leading to poor heat transfer
  - Every 2-4" is recommended
- Installation in an electrically classified area is very difficult
  - Exterior surface almost always exceeds autoignition temperatures no matter how low the control temperature
  - Any leaks will have to pass an ignition source

11

ExxonMobil
Research and Engineering

Potential Solutions

- Purchasing higher voltage tape and running it at lower voltage (e.g. 120 VAC on a 240 VAC tape)
  - Often more successful due to averaging effect of heater sheath and internal fill
  - Very dependent on actual installation so testing after installation is required
  - Be careful that additional insulation will not inadvertently lead to higher temperatures
- Limiting the voltage (variable transformer) or amperage (phase angle fired SCR)
  - Many SCR's cannot limit either and the heater wire will reach operating temperatures but for a shorter time which is not safe
  - Again testing is recommended

12

ExxonMobil
Research and Engineering

Potential Solutions

- Putting a thermocouple on each tubular heater and controlling to below the autoignition temperature
  - Requires separate control and alarm thermocouples and controllers
  - Finding the hottest point on the heater may require testing after installation

Tubular Resistance Heaters

- Tubular heaters are difficult to wrap around supports, instruments and other heat sinks so cold spots are more likely
  - Care necessary in piping design
    - + Insulation breaks
    - + Supports around insulation instead of piping
  - The first application is usually the best so advance planning and care in installation is more effective than trying to correct problems
- Burn hazards are very common
  - Often difficult to identify in advance
  - Usually found only after a burn is received
  - Extreme care in initial testing and operation is required
  - Thermal imaging devices can be very helpful in identifying

Tubular Resistance Heaters

- Heat transfer cement can be effective
  - But only if the heater's thermal expansion of the heater is controlled or the cement usually has a short life due to breaking off
- Heat transfer tapes usually are marginal
  - Most have difficulty due to high heat fluxes
  - Many begin to separate from heaters over time reducing effectiveness

15

ExxonMobil
Research and Engineering

Steam and Hot Oil

- Modifications are costly
  - Plan for extra zones
  - Deliberately undersize zone lengths so additions are possible
- Restrict zone lengths to reasonable runs and not just available pressure drop
  - Long zones often lead to cool area at the end when all the steam has condensed or the hot oil's ΔT is almost exhausted
  - Hydraulic problems can arise
- Leakage is always a potential problem
  - Keep joints outside the main insulation so they can be more easily accessible for inspection and repair
  - Weld or braze connections instead of using fittings to minimize leakage

16

ExxonMobil
Research and Engineering

Steam and Hot Oil

- Heat transfer cements can improve effectiveness
  - Particularly noticeable on larger components (vessels, tanks, etc.)
  - Less effective on smaller piping
- Consider the use of jacketed lines instead of tracing
  - Significantly more effective due to excellent heat transfer area
  - Fittings need to be bypassed
  - Most cost effective for longer runs with few or no fittings

Steam

- Always provide a dedicated steam trap for each zone
  - Ganging zones on a common trap always leads to some areas not working properly
- Recognize that steam traps will require some routine maintenance
  - A sudden "improvement" in a zone is often an indication the trap is bypassing live steam

ExxonMobil
Research and Engineering

Hot Oil

- Most heating fluids break down over time
  - Consult the manufacturer's recommendations for testing/replacement timing
- Many high temperature fluids are or become hazardous
  - Evaluate how to safely change out degraded fluids before construction
  - It may require additional piping and/or safeguards (ventilation, closed transfers, etc.)

Heated Enclosures

- Easy to modify process as nothing is affixed to the piping
  - Difficult → impossible to retrofit so evaluate carefully before construction
- Commercial strip heaters provide longer life and lower cost
- Provide air circulation to minimize temperature differentials
  - Simply relying on heat rising by putting the heaters on the base of the box often effective for small enclosures
  - Placing inlet holes/strips at the base and outlet hole/strips at the top often effective for moderate size and/or lower temperature enclosures
  - Air amplifiers (higher temperatures) or fan (lower temperatures) are required for larger enclosures

Heated Enclosures

- Evaluate how valves and similar components will be actuated
  - Requires additional care in layout to make sure all components are accessible from outside
  - Occasionally opening a door to turn a valve may sometimes be acceptable
  - Extending valve handles is normally the preferred solution
    + Some additional extension for cooling may be required
    + Be careful that the attachment mechanism cannot loosen with repeated heating/cooling
  - Remote actuation may sometimes be required
- Meeting area electrical classification is relatively straightforward

21

ExxonMobil
Research and Engineering

Potential Solutions

- Purchasing higher voltage heater and running it at lower voltage
  - Testing before use is recommended but usually more successful due to averaging effect of heater sheath and internal fill
- Limiting the voltage (variable transformer) or amperage (phase angle fired SCR)
  - Many SCR's cannot limit either and the heater will reach operating temperatures but for a shorter time which is not safe
  - Again testing is recommended
- Purging per NFPA-496
  - Allows temperatures above the autoignition temperature
  - Inert gas purging may be required
- Limiting the surface temperature of the heater

22

ExxonMobil
Research and Engineering

Surface Temperature Limitation

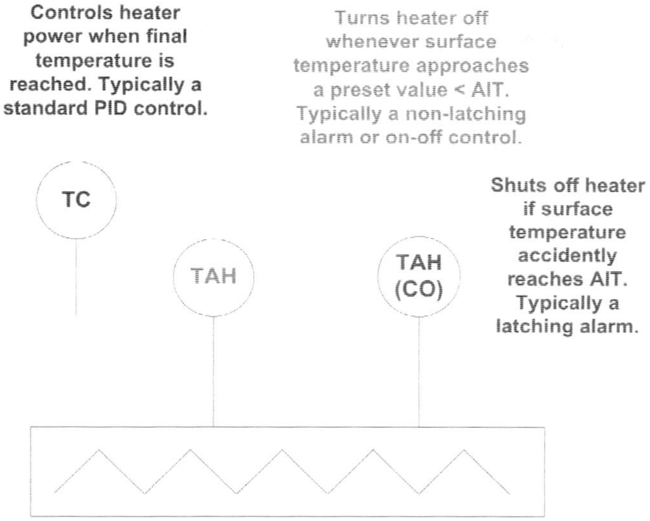

Caution

•The power delivered will be reduced proportional to the reduction in surface temperature.

•Many heaters will tend to overshoot significantly past TAH so a larger offset from TAH(CO) may be required.

Heated Enclosures

- Burn hazards while significantly less of a problem can still arise in unexpected places
  - Common problem areas
    - + Doors and panel clearances
    - + Valve handle extensions immediately next to the heated enclosure
    - + Process lines immediately next to the heated enclosure
    - + Structural supports
  - Usually easily found but some care in initial operation is required
  - Thermal imaging devices can be very helpful in identifying

Less Effective Approaches

- Preheaters
  - Usually only viable if the heated section is very short like a reactor
- Heat lamps
  - Limited to low temperatures and general purpose areas
- Forced hot air systems
  - Limited to very low temperatures or system complexity increases significantly

25

ExxonMobil
Research and Engineering

Safety

- Make sure the method you are considering can meet the required area electrical classification
  - Self limiting and steam are generally inherently safe
  - Constant wattage and tubular resistance tracing require careful over temperature protection
  - Heated enclosures are recommended for higher temperatures
- Expansion and contraction following heating and cooling can lead to routine process leakage
  - Consider eliminating joints by welding or single lengths of tubing
  - Determine if keeping the pilot plant hot all the time is required
  - Make sure routine leak testing is a part of the start up procedure and is well thought out and carefully done

26

ExxonMobil
Research and Engineering

Safety

- Personnel protection needs consideration
  - Maintaining proper insulation requires routine attention
  - Proper wiring and grounding is critical
  - Lockout/tag out provisions require enforcement
  - All electric tracing requires GFI protection by the NEC

Design

- Be realistic about how to provide over temperature protection
  - Almost always requires post construction testing (and modification!)
- Be skeptical of published heat loss values based on larger installations, components or piping
  - Higher density of valves and supports often lead to much higher values
  - Insulation is rarely as well done
- Consider the need for standoffs to isolate heated components from supports
- Insulation installation is as important as the heating yet is often very poorly done

Operations

- Carefully review the need for access to the process piping
  - Lock out and tag provisions
  - Removal and reattachment issues
  - Re-insulation
- Consider the need for additional temperature indication to confirm proper operation

29

ExxonMobil
Research and Engineering

Electric Tracing

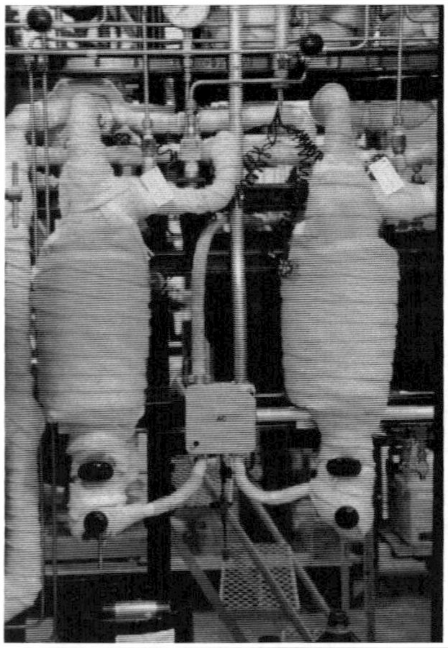

30

ExxonMobil
Research and Engineering

Steam Tracing

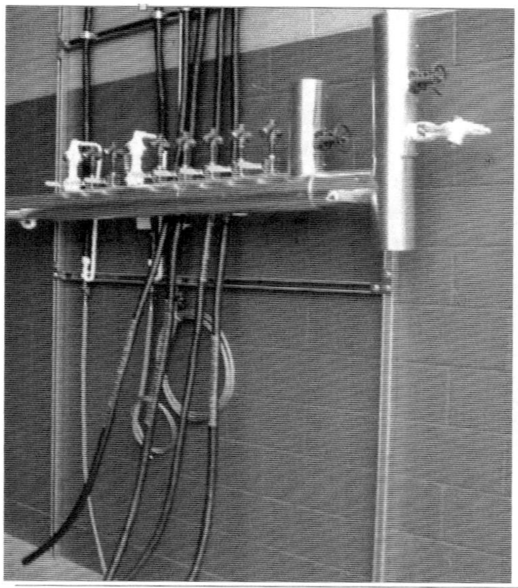

Heated Enclosures

(115c) Pilot Plant Best Practices for Safer Design and Operation

Kiran Gupte, BP Refining & Logistics Technology, Naperville, IL

BP Refining & Logistics Technology group has successfully designed, commissioned and operated numerous pilot plants in support of refining processes over the last few years. However, many of the practices adopted by the team for safe and reliable operation have been learned over more than 20 years of pilot plant experience at Naperville site. Of course, over the years the management practices have evolved with newer methodologies. In addition to the old and reliable "Process Safety Review (PSR)" process, the newer "Management of Change (MOC)" procedures have been implemented to further strengthen the safe operations. An "Inherently Safer Design" is at the heart of a new pilot plant during the conceptual design phase. The importance of "Design Phase" can not be over emphasized from the cost perspective to eliminate any cost overruns due to changes that may be needed to correct undersized or wrong equipment. This presentation will also highlight the challenges of designing and operating small scale equipment that are needed at the pilot plant scale. The pilot plant designers, operators & engineers need to appreciate the differences between pilot plant and commercial scale in order to fine tune the standard practices for safer design and operation. The attention to "Operating within the Design Envelope" while utilizing Task Hazard Risk Evaluations and Mitigations is important for safe operation with zero incidents. All these aspects are an integral part of the overarching Operations Management System. The safe practices developed are applicable to all types of equipment including small-scale catalyst screening reactors, small-medium autoclave reactors, distillation apparatus, and large pilot plant and demonstration reactors handling a broad range of hydrocarbon feedstocks.

Mini- and Micro-Scale Units in Process Development

Joseph B. Powell, Ph.D.
Shell Chief Scientist – Chemical Engineering, Houston, TX

Introduction

As bench scale and pilot units increase in size, safety risks from fire, explosion, or toxic exposure increase, as does overall footprint, costs, and required staffing. Time required for implementation increases dramatically upon progression from micro scale screening, to conventional pilot scale operation. A key question in process development is the minimum scale needed to reliably obtain information needed to assess process viability. A second critical assessment is to determine what scales and steps for pilot and demonstration facilities would be needed, if the process is found to be viable, in order to de-risk a commercial investment?

Quite often, the optimal answer to these questions entails one or more readily reconfigurable mini- or micro-scale units to allow rapid assessment of viability of key process steps, and some degree of integration between steps. Often, a further advantage of the micro- or mini-scale units is the ability to build and run multiple tests in parallel, which speeds the rate of determining the optimal process concept. Optimally, a larger scale pilot can then be designed and constructed to provide the necessary information for scale up to commercial, avoiding an intermediate demonstration phase. By using the micro- and mini-scale units to optimize among process options, more expensive and time consuming process changes at the larger pilot scale can be minimized.

Mini- and micro-scale experiments [1-5] often suffer however from less precision, and integration between units often must be done via simulation, with calculated results used to prepare mock fluids of anticipated steady state compositions as feeds. Reduced precision is typically best accommodated via use of side-by-side or consecutive comparison with known systems as a benchmark. Contacting history and time constants for all rate-dependent mechanisms will in general not reproduce commercial operation, but underlying phenomenon can be isolated, magnified, and studied, with appropriate experimental design [5]. Significant time is required in design of experiments, in order to draw out, understand, and magnify the desired parameter effects, while also eliminating possible anomalies from operation at the micro- or mini-scale.

Laboratory Catalytic Reactors

Catalytic fixed-bed reactors have historically been sized via minimum tube diameter to prevent channeling [6], which for commercially-relevant catalysts typically means one-inch diameter or larger tubes packed with 10 – 100 grams of catalyst. Footprint and required throughputs for these bench scale or mini-pilot units is relatively large. In recent years, use of catalyst particles immersed in finely divided diluent [7] has been established as a viable method for screening these systems, especially for hydrocarbon solvents, and especially where side-by-side or sequential comparison testing is conducted.

Bubble columns or stirred reactors are conveniently studied via laboratory stirred reactors, where catalyst baskets can be used to prevent fixed-bed catalyst attrition. Scale-up from stirred laboratory reactor to commercial bubble column can be done from first principles, without having to re-pilot bubble column performance. Similar scale-up from stirred reactor to commercial loop reactor can be anticipated as feasible, if potential for erosion in presence of suspended solids is anticipated.

Separations

Relatively less attention has been given to scale-down of separations [1]. Often, a crude separation step is needed during process development to provide *integration* between process steps including multi-step reaction processes. Adsorption, ion exchange, or even crystallization can be conducted at the micro scale via use of tumbling or stirred vials [1]. Simulation of thermal separations (distillation) at the micro scale is more difficult, unless a simple single-stage flash is involved. "Mini" scale batch distillations can be conducted to determine separation factors relative to known components, to calibrate interaction parameters needed for numerical simulation. One alternate approach is to not do the separation, but to use analytical characterizations and kinetic modeling to examine the integration between first and second step reactions.

Recent success was obtained in implementing micro-scale separation via use of sintered metal filtration to assess rates of dissolution, and eliminate plugging of fixed-bed catalyst, for a micro-scale coupled pilot unit. Periodic manual blowback to remove entrained solids provided both analytical information on dissolution kinetics from step-1, and provided viability for directly coupled fixed-bed reaction in step-2. Use of oversized filters allowed blowback frequency to be set at 1.5 hours vs. anticipated 15 minutes for commercial operation, in order to avoid time delay required for implementation of automated back flush, for a system involving liquid flows at 0.02 to 0.20 ml/min.

Virtual Integration Example

Figure 1 presents an example result from a study where micro-scale batch reactors were used to assess the impact of liquid phase residence times on process yields for a two-step continuous flow process, where a separate fixed-bed catalyst would be employed for the second conversion step. For the commercial process, steps #1 and #2 would be coupled in series with no intermediate cooling. A batch experimental sequence with crushed catalyst (slurry form, less than 100 microns) was employed, to test the impact of separate step-1 residence time and test for possible anomaly due to cool down required for step-2 catalyst addition. Results quickly showed a negative impact on yields from step-1 residence time, with a minor impact of the laboratory anomaly associated with cool down interval between steps. The result directed efforts to development of alternate integrated process concepts that did not require sequential two-step contacting, for maximization of yields. Reactor type and contacting history did not match proposed commercial units, but by careful design of experiments, process options could be screened in much less time than would have been required for assembly of a two-step, continuous flow integrated pilot.

Importance of Side-by-Side Comparison

A second micro-scale pilot example entailed ion exchange pretreatment, reaction, separation by crystallization, and a final thermal finishing step. Anomalous behavior in the corresponding commercial unit was diagnosed by doping impurities with feed to the initial pretreatment step. None of the separation factors or kinetics and reaction profiles matched the continuous commercial process, and only the thermal heating (and not separation) was implemented for the final thermal separation step, with analytical assessment of result. Side-by-side comparison across a sequence of batch reaction and separation steps with analytical characterization of the resulting product yields, where each step employed less than 10 grams of material, was successful in rapidly diagnosing the source of trace chemistry responsible for off-spec commercial product. Final isolated product was never produced in the lab sequence.

Over-instrumentation trap

There is a natural tendency to over engineer new experimental equipment, especially at the design phase. In one recent example, two new batch multireactor units were purchased for separate process development programs. Catalyst and reactants were manually charged to Unit A, which was also manually sampled for liquid phase GC analysis. The latter did involve automation in the form of a standard autosampler ("robot"). Unit B cost more than 5X more, and was set up for automated charge of reactants, and automated liquid phase sampling with GC analysis.

More than 1500 experiments were completed in Unit A, comprising the entire extent of study needed for project completion, before the first automated experiment was completed in Unit B. A third study added a catalytic pulse microreactor as the insert of on a GC equipped with autosampler, such that reactants could be varied by simply manually changing the composition of feeds to autosampler vials. This unit produced results within one month of concept.

The lesson from these examples is that new automation can come at a significant cost in time and instrumentation. If the experiments will be long term repetitive across multiple projects, then automation can be warranted. Process development often involves streaming of different types of experiments and equipment for different projects, and hence the leverage potentially achieved from automation is often offset by changes in process and apparatus configuration.

Summary

Micro- and mini-scale experiments can often be coupled to simulate the rudimentary aspects of integrated processes, where thermal stresses and reaction times can be studied, even in the absence of precise separation or residence time history. Careful use of side-by-side experimentation and comparison with" known" behaviors validated in larger scale pilot or commercial facilities, can be used to diagnose impacts or effects of changes in process variables.

Mini- and micro-scale studies can thus be used in tandem with larger scale pilot operations, as a means for parallel study of process variables and impacts, using the larger pilot as calibration. Impacts of individual phenomenon can often be magnified to clearly measurable levels (e.g. impacts of potential catalyst poisons), which are not evident due to noise in pilot plant studies locked in to operation at design conditions.

References

1. J. B. Powell , *Shortcut Methods for Scale-up of Separation Processes*, AIChE Spring National Meeting, Tampa, FL, April 26-30, 2009
2. J. B. Powell ,*Scale-down Considerations for Pilot Plants and Bench-Scale Units,* **AIChE Spring National Meeting***, New Orleans, LA, April 6 – 11, 2008.*
3. J. B. Powell , *Reactor Scale-down for Pilot Plant, Bench Scale, and Multi-Throughput Units***, AIChE 2006 Fall National Meeting**, November 12 – 17, San Francisco, CA.
4. J. B. Powell, *Use of Shortcut Methods in Process Development*, **AIChE Fall Annual Meeting**, Cincinnati, OH, Oct. 30 – Nov. 4 (2005
5. J. B. Powell , *PROCESS DEVELOPMENT: IS SCALE-UP REALLY ABOUT SCALE?, 8th World Congress of Chemical Engineering, Montreal, Canada, August 23rd-27th, 2009.*
6. D. E. Mears, "Transport Effects in Laboratory Catalytic Reactors", *Ind. Eng. Chem. Proc. Des. Dev., 10*, 541 (1971).
7. S. K. Bej, R. P. Dabral, P. C. Gupta, K. K. Mittal, G. S. Sen,V. K. Kapoor, and A. K. Dalai, *Studies on the Performance of a Microscale Trickle Bed Reactor Using Different Sizes of Diluent*, **Energy & Fuels, 14 (2000)**, 701-705
8. P. M. Wilkinson, A. P. Spek, and L. L. van Dierendonck, "Design Parameters for Estimation of Scale-Up of High Pressure Bubble Columns," *AIChE J., 38*, 544-554 (1992).

Figure 1: Parallel Batch Slurry Catalyst Experiments for study of Sequential Fixed-Bed Process Concept. [Test impact of separate step-1 residence time, parametric in laboratory-required cooling interval between steps].

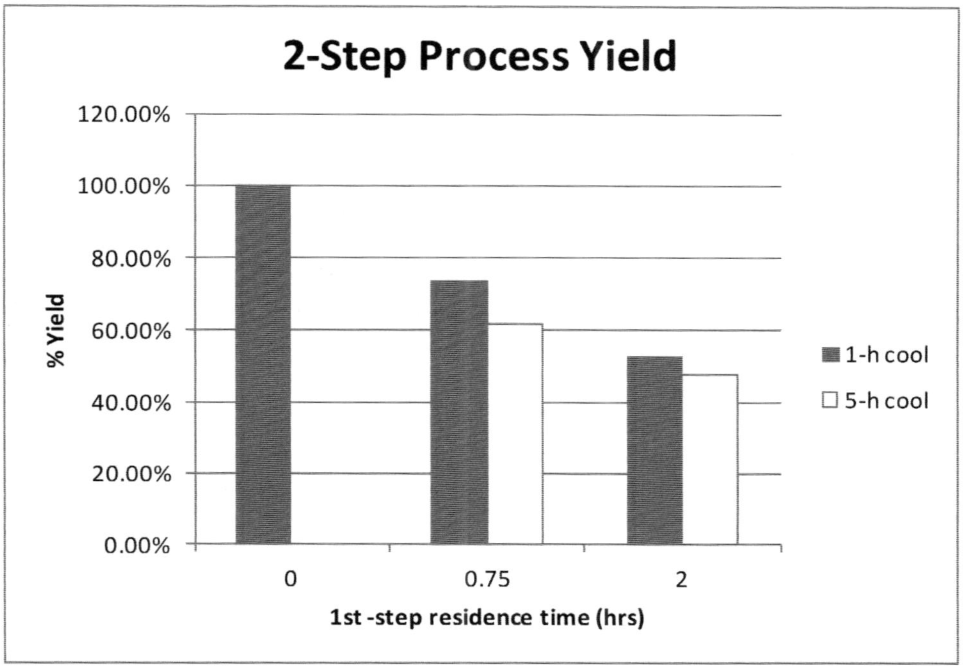

Data Integration in Petrochemical Pilot Plants by Software Optimization

Hsiang Yee Lai, Chemistry and Chemical Engineering Division, Southwest Research Institute
Theodore Wilmes, Automation and Data Systems Division, Southwest Research Institute

Introduction

Petrochemical pilot plants differ from commercial plants in that they are not continuously in steady-state operation. Petrochemical pilot plants consistently experience startups and shutdowns to allow for process design modifications. Process conditions in pilot plants vary depending on specific experimental parameters. A large amount of analytical and pilot plant process data is accumulated over time. Therefore, software development to efficiently integrate analytical and pilot plant process data becomes critical to petrochemical pilot plant operations.

The core of process engineering in pilot plants relies on the accuracy and consistency of process and analytical data acquisition. Chemical engineers incorporate this data to build multivariate statistical and thermodynamic models, which may predict methods to optimize yields and product selectivity in a pilot plant. Additionally, this data interpretation may result in system design modifications that may further improve the product characterization and overall efficiency of the system. The integration of process and analytical data into a centralized location is essential to prepare for data interpretation.

With the rise of computing technology and the increased use of process control management systems, pilot plant process data acquisition has become increasingly automated. Process data can be acquired more reliably and at a higher frequency. Several commercially available process control management software packages store historical pilot plant process data in a database format.

Analytical data, however, is not acquired through a management system. Process samples are taken by pilot plant operators and sent to chemical laboratories for analysis. A data engineer will provide a list of requested analyses for the samples. Chemists and chemical technicians analyze the samples and report the results to the data engineer. Analytical data may be received from different sources in varying formats (e.g. spreadsheets, emails, and word documents), thus increasing the possibility of communication and transcription errors. If a Laboratory Information Management System (LIMS) is in place, analytical data flow between chemical analysts and data engineers is streamlined. Data will arrive in a standard format from a single source, and is typically not stored in the same location as the pilot plant process data.

The disparity between the process and analytical data formats raises obstacles for the preparation and centralization of combined data. Data engineers end up spending more time compiling data compared to analyzing and interpreting data. Consequently, design modifications and yield optimization resulting from data analysis requires a longer turnaround time due to more time spent centralizing the data. Data flow becomes congested between pilot plant operations, chemical analysts, and data engineers. Figure 1below highlights the flow of data between each entity for a currently operational petrochemical pilot plant.

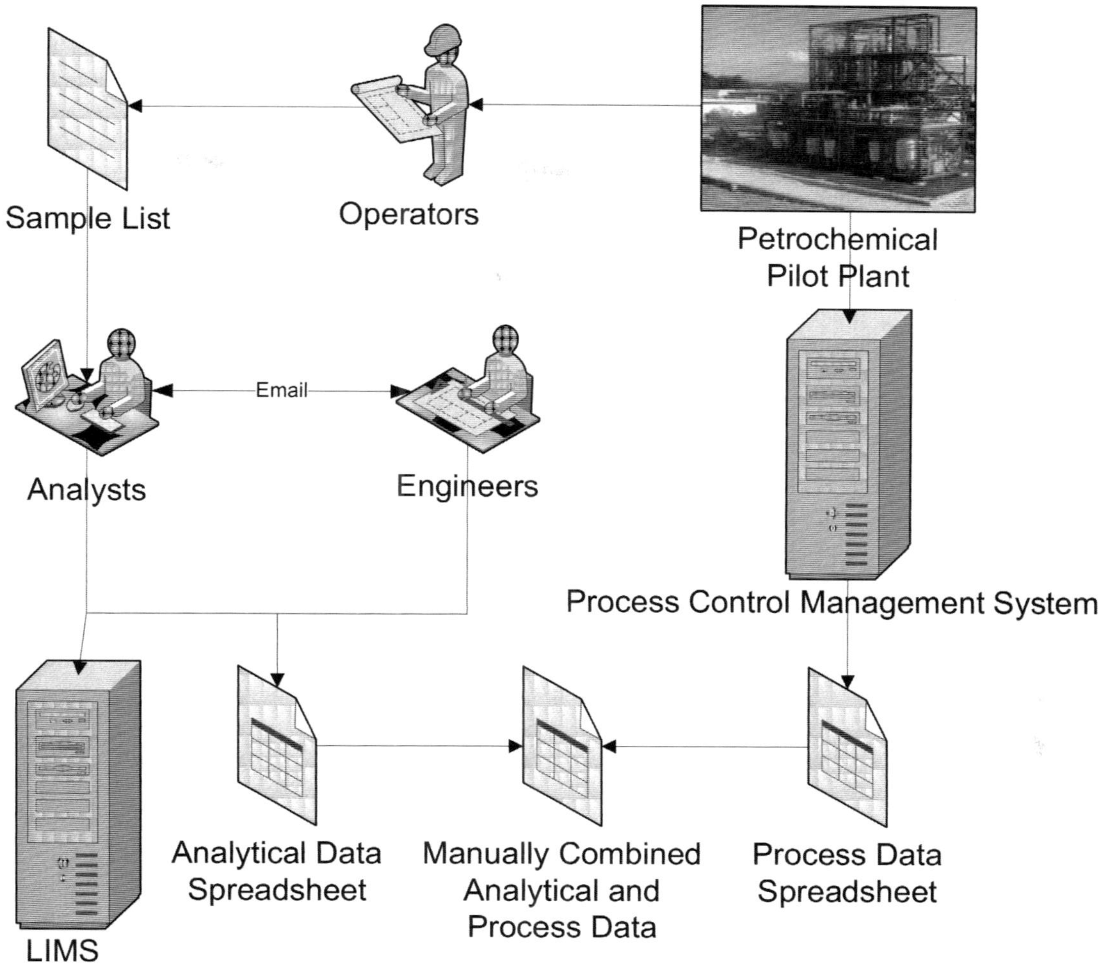

Figure 1: Original Pilot Plant Data Flow

A suggested solution to this dilemma is business process optimization. The purpose of business process optimization is to automate tasks with software where possible allowing personnel to focus on their specialties. In order to centralize and streamline the data flow between the pilot plant, chemical analysts, and data analysts, a data warehouse and customized software system were integrated into an operational petrochemical pilot plant.

Three steps were taken to achieve this: streamlining sample documentation, analytical data access, process data access.

Streamlining Sample Documentation

The existing sample documentation process was identified as an area where changes could be made to improve efficiency. Data engineer and analytical staff time was consumed by duplicative spreadsheet management and data entry. As a result, multiple points of failure were introduced where data could be entered incorrectly or simply lost, thus impacting data integrity.

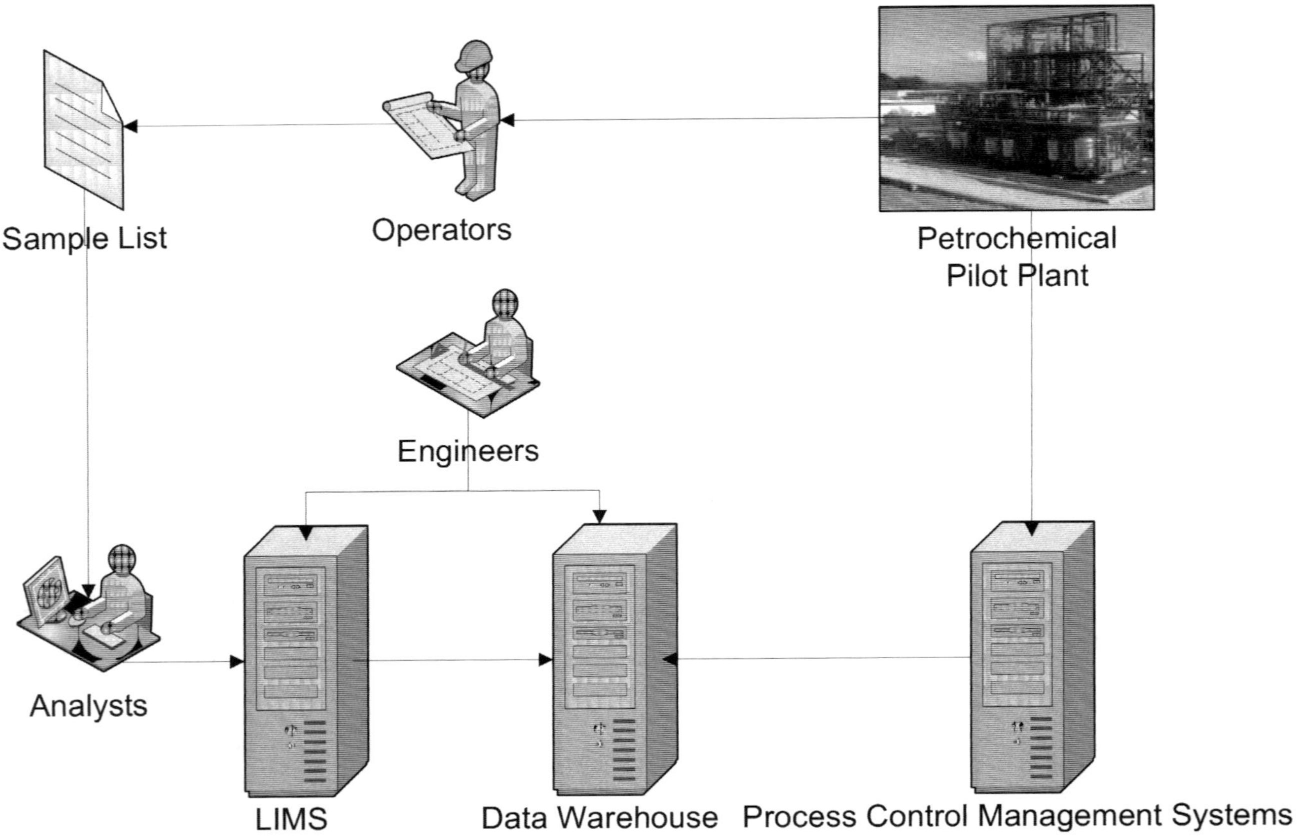

Figure 2: Data Warehouse Centered Process

The first step of the sample documentation process involved the collection of samples by pilot plant operators. Engineers would then identify and document these samples in a spreadsheet. Analytical requests would also be saved to the same spreadsheet. The samples would then be delivered to the chemical laboratories. Chemical analysts would enter the sample information into the available LIMS. The purpose was solely for sample tracking purposes. Once tests were completed, analysts would enter the test results directly into a set of shared spreadsheets. There was minimal auditing capability to track data entry, which allowed for transcription errors and a lack of accountability for said errors.

At the outset of the data integration effort, project personnel found the existing LIMS could be used to document test assignments and results. Subsequently, the LIMS was used as the analytical data source for the data warehouse.

The new process, illustrated in Figure 2, moved sample and test result entry into the LIMS. Initially, pilot plant operators collect samples, which are then picked up and documented into the LIMS by the analysts. Engineers then create task orders in the LIMS, requesting specific testing analysis for samples. Upon completion of a test analysis, the chemical analyst enters the data directly into the LIMS. The data entered into the LIMS is stored in a database on a network. Software development is required to integrate the LIMS database into a data warehouse.

The data warehouse serves as a centralized repository for the analytical and process data. Software was developed to extract analytical data from the LIMS directly into the data warehouse.

Because the LIMS data is mostly unstructured, it is first stored in a temporary holding area. An automated process parses the sample ID into discrete parts such as run, sample location, and sample number. Since sample IDs are not machine generated, the algorithm does not always recognize the sample format. As a result, sample documentation anomalies are identified. This normalization allows for query flexibility further down the road.

If the system cannot parse the sample ID, the user must remediate the issue. A custom graphical client is used for this task. The user will create the correct sample ID and then link the data of the unrecognized sample to the corrected sample identity. From this point forward, test results for this sample will be automatically recognized by the system and imported without user intervention. This step is necessary only because the LIMS does not have support for discrete sample IDs that matched the pilot plant project requirements for tracking.

Analytical Data Access

As data flows into the data warehouse, it becomes available for the engineers and other interested personnel. A custom reporting tool has been developed that simplifies report development. One of the biggest benefits of the data warehouse over just a pure LIMS solution is that the data warehouse can hold pilot plant specific information in a manner that allows for dynamic retrieval and filtering of the test data. For example, pieces of information that could only be stored as free text in the LIMS are stored in a structured format in the data warehouse. This allows system users to retrieve samples and test results based upon criteria that make the most sense to their daily work with the pilot plant. Each sample is associated with a run, location in the pilot plant, matter state of the sample (solid, liquid, gas), a unique id number, and date and time the sample was taken. The reporting tool provides the capability to filter on all of these attributes.

The custom reporting tool is not meant to replace spreadsheets or other off the shelf reporting tools, but to act as an intermediary between the data warehouse and the user. It quickly and efficiently retrieves analytical data from the data warehouse in a user-specified report format. Users first select the analytical tests that they are interested in retrieving results for and then select the samples of interest based upon specified filtering criteria. From this, the user can generate a report. The report is displayed in a tabular format similar to a spreadsheet. Data can be studied directly in the tool, or the table can be stored in spreadsheet format. In many cases, reports for outside sources still operate on spreadsheet files so this provides the necessary backwards compatibility.

Process Data Access

The industrial process control management system serves as the process data source for the data warehouse. It is important to store process data in the data warehouse to easily facilitate analyses that require an insight into analytical and process data to improve pilot plant operations. Traditionally, process data was extracted from the database in a spreadsheet format. The highest granularity of data extracted with spreadsheets was at a frequency of one minute. This would require two hours of a data engineer's attention to achieve on a weekly basis. The modified system automatically retrieves larger, bulk process data extracts, and loads them into the data warehouse. Data is extracted from the process control management system at the highest level of granularity available, which for some process variables is one read per second. As part of the import process, the process data is also stored at a one minute average granularity. This allows for greatly improved retrieval speeds and for many cases, is more than adequate resolution.

The process data is accessed primarily through a custom developed charting tool. As with the analytical client, meta-data is present in the database that allows the process data to be queried in a manner that makes sense for the pilot plant. For example, users can retrieve process data by pilot plant run as opposed to having to know the specific date range the run occurred. In addition to plotting process data, the process data client allows users to overlay analytical test results on the same charts. Calculated or derived process variables can also be computed and charted. For example, engineers can enter an equation into the system that makes use of concrete process variables and then chart the result.

Advantages and Disadvantages of Software Development for Data Integration

The overall advantages to this system are the centralization of the analytical and pilot plant process data into a data warehouse and the re-allocation of time spent compiling and preparing data for further analysis.

Previously, engineers received analytical testing results from chemical analysts in multiple formats (e.g. cumbersome shared spreadsheets, emails, pdfs). Engineers were expected to accurately compile analytical data and produce data reports to the client. Within this system, the same analytical data would be transcribed multiple times, introducing many possibilities for human error without a systematic method to audit errors.

Additionally, pilot plant process data had to be transferred from a process control management system database to large spreadsheet files by an engineer. Subsequently, these files needed to be loaded to the network and modified for further analyses purposes.

With the advent of the newly developed system, analytical data entry errors became traceable and easily remediated. Time spent organizing and preparing analytical and pilot plant process data has been reduced. Engineers are now able to concentrate on engineer-centric tasks such as analytical data interpretation and multivariate analyses of process variables. As a result, pilot plant improvements and final petroleum product development have moved at a faster pace.

The pitfall of initiating software development to optimize process engineering is to not clearly define the objective and purpose of the software. By avoiding this pitfall, the software development is more likely to be productive. Initially, the development of the data warehouse and software tools requires a large time and cost investment. This allows the engineer, analysts, and operations personnel to effectively improve the pilot plant system to produce the desired products and yields.

Future Development

Future development of the pilot plant data integration solution will extend the analytical capabilities from the first level of data retrieval to greater automation of data analysis using data mining techniques. The centralization of process and analytical data in the data warehouse simplifies the task of finding correlations between process conditions and process output quality. The primary interest is to predict process output based upon conceptual process conditions or hypothetical feedstocks. Other applications of data mining include root case analysis. Time series process data can be compared run to run to determine how process conditions may have led to degraded plant output.

AUTHOR INDEX